PLANTS THAT
Kill

PLANTS THAT
Kill

A Natural History of the World's
Most Poisonous Plants

Elizabeth A. Dauncey
and Sonny Larsson

PRINCETON UNIVERSITY PRESS
PRINCETON AND OXFORD

This edition published in North America in 2018 by Princeton University Press,
41 William Street, Princeton, New Jersey 08540
press.princeton.edu

© 2018 Quarto Publishing plc

All Rights Reserved

ISBN: 978-0-691-17876-9

Library of Congress Control Number: 2017952293

British Library Cataloging-in-Publication Data is available

This book has been composed in Monkton and Frutiger
Printed on acid-free paper

This book was conceived, designed, and produced by Global Book Publishing,
an imprint of The Quarto Group

Written by Elizabeth A. Dauncey and Sonny Larsson
Additional text by Sarah E. Edwards and Kathryn Harkup
Designed by Lindsey Johns
Project managed by D & N Publishing, Wiltshire
Commissioned by Jacqui Sayers

Photograph opposite: red angel's trumpet (*Brugmansia sanguinea*)

Printed in China

10 9 8 7 6 5 4 3 2 1

Disclaimer

The information in this book is intended to educate, delight and expand the reader's understanding of the diversity of plant life, the compounds they produce and their effects on animals and humans in particular. It is for these reasons that the authors and publishers have made this information available to the public. The authors and publishers do not endorse the use of these plants for any of the applications described and are not responsible for any consequences arising from the use of this information for whatever reason, including curiosity or malicious or illegal intent. The absence of a plant from this book does not necessarily mean that it is non-toxic. It is not intended that this information should be used to diagnose poisoning. If poisoning is suspected, medical or veterinary attention should be sought immediately and a piece of the plant taken with the patient to aid diagnosis.

Contents

INTRODUCTION 8

CHAPTER 1
WHY SOME PLANTS ARE TOXIC 10

CHAPTER 2
TARGETS IN THE BODY 28

CHAPTER 3
MATTERS OF THE HEART 44

CHAPTER 4
BREAKING THE BRAIN 62

CHAPTER 5
MORE THAN A WEAKNESS OF THE KNEES 90

CHAPTER 6
STARTING AS AN IRRITATION 106

CHAPTER 7
GUTS WITHOUT THE GLORY 132

CHAPTER 8
ORGAN FAILURE 154

CHAPTER 9
CELL POISONS 176

CHAPTER 10
TURNING FOES INTO FRIENDS 194

GLOSSARY 218

FURTHER READING 219

INDEX 220

ACKNOWLEDGEMENTS AND PICTURE CREDITS 224

Introduction

Plants produce poisons to protect themselves, sometimes at levels toxic enough to kill. Through trial and error, our forebears established which plants were edible and which caused harm. They then avoided the harmful species or in some cases also knowingly used them as weapons to kill rivals, criminals and animals, or as drugs to evoke hallucinations or subject victims to trial by ordeal. While knowledge of poisonous plants may have been lost among the majority of us who now buy our food in shops, scientific understanding of plant toxins and their effects has increased greatly.

Scope

This book uses a combination of text, photographs, diagrams and chemical structures to create a fascinating picture of how and why plants – mainly flowering plants – produce poisons (also known as toxins). Taking a global view, it describes a selection of the most historically or culturally significant, interesting and important poisonous plants, the compounds they produce and the mechanisms through which they work on animals – humans in particular. The latest research has been used throughout and is presented in a readily accessible way.

There are, of course, different degrees of toxicity, with some plants just making us a little sick. The plants described in this book, however, are at the more serious end of the spectrum; in fact, they are known to kill – hence the title of this book. Plants that cause serious effects following contact are also included. While the symptoms resulting from such contact usually fall short of actual death in larger animals, they can be very debilitating, and often these plants are lethal to smaller animals or microorganisms.

Outside the scope of this book are carnivorous and parasitic plants. Most fungi are also excluded, the exception being a few fungi that exert their effects through their close association with plants.

BELOW **Stems of oleander (*Nerium oleander*), a widely cultivated shrub from the Mediterranean, may have poisoned soldiers serving under Alexander the Great when used by them as meat skewers.**

How this book works

Title
The type of compound or plant covered on the pages, sometimes also with an indication of the effects.

Chemical structures
The structures of the main toxic compounds found in the featured plant or plants. Provided to show the variety of structures that plants produce and to enable readers to make comparisons between compounds.

Images
Usually photographs to illustrate the plant or particular parts of the plant; botanical drawings or coloured illustrations are used on family spreads, in Chapter 10 (which looks at medicinal applications of poisonous plants) and also occasionally on standard spreads.

Yew and your heart

Taxus was the Latin name used by Romans for the yew tree, so the choice by Carl Linnaeus of the same name for the toxic genus seems pretty obvious. However, the etymology of the word is particularly interesting in a book about poisonous plants. The Romans are believed to have taken the name from the Greeks, who constructed the word *toxikon*, meaning a poison or drug used on arrows, from their word for a bow (*toxon*). As yew wood has a reputation of being the best for making bows, we have, in a way, come full circle.

Deadly needles

PLANT:
Taxus baccata L.
COMMON NAMES:
yew, English yew, European yew
FAMILY:
yew (Taxaceae)
TYPE OF TOXIN:
Taxus alkaloids (taxine B)

SYMPTOMS OF POISONING IN HUMANS:
CIRCULATORY: abnormal heartbeat
NEUROLOGICAL: dilation of the pupils, dizziness, weakness, coma
DIGESTIVE: abdominal cramping, vomiting

ABOVE Taxine B is a diterpene alkaloid due to the presence of a nitrogen atom in its sidechain. It is found in many species of yew (*Taxus*) and is structurally similar to the cancer drug paclitaxel.

BELOW Underside of a yew (*Taxus baccata*) branch, showing the pale lower surface of its needle-shaped leaves and both immature green and mature red berry-like arils, each surrounding a single seed.

The yew is one of the few plants in this book that is not a flowering plant, but rather a conifer. As is typical of most other conifers, yews are trees or shrubs with modified leaves that we call 'needles'. Their seed 'cone', however, is very different from a pine cone: it is single-seeded, and by the time each seed is mature it is surrounded by a cup-shaped berry-like aril.

Yews are very slow-growing and long-lived. Probably the oldest living yew tree in Europe can be found in St Cynog's churchyard in Defynnog, Wales, and is thought to be around 5,000 years old. There is also evidence that the genus itself is very old in evolutionary terms. *Paleotaxus rediviva* fossils from the Triassic, dating back some 200 million years, are recognizable as yews: as are those of *Taxus jurassica* from the mid-Jurassic, 140 million years ago.

Today, the *Taxus* genus includes 12 species found around the world, including much of Europe, north Africa, China, the Philippines and Sumatra, Mexico, the United States and Canada.

All parts of the yew, with the exception of the arils, contain taxine alkaloids. The toxicity is not decreased on drying, so hedge trimmings are as toxic as the plant itself. Some deer seem able to eat the foliage, and sheep have been known to browse on it unharmed. However, other animals – including horses, cattle, dogs and humans – are poisoned if they eat the leaves or branches. Poisoning of livestock is usually prevented by ensuring that they do not have access to these trees.

Food for badgers
The non-toxic arils produced by yews are sweet and of a gelatinous consistency when ripe. Their scarlet or scarlet-orange colour makes them easily seen by birds, which eat them whole, including the seed. The seed then passes through the bird and is deposited elsewhere. Children attracted by the arils usually have the good sense to spit out the seeds, which are larger than the seeds of a grape, and so do not come to any harm.

Scientists at the Royal Botanic Gardens, Kew, observed European badgers (*Meles meles*) eating yew arils off the ground and even standing on their back legs to eat them off the trees. They wondered why the animals were not poisoned. Dung pits found nearby that were full of partially digested arils and apparently intact seeds provided a possible answer. But being scientists, they wanted to test whether the seeds were in fact undamaged by the badgers' digestive system. They used liquid-chromatography mass-spectrometry (LC-MS) analysis (which separates and measures the mass or weight of compounds) to determine the presence and concentrations of alkaloids in arils and seeds taken from the trees, and also in seeds that had passed through the badgers. They found that there was no difference in the concentration of the main toxins in the seeds before or after they had been eaten. They also confirmed that the arils contained no alkaloids. The appearance of the arils in the dung suggests that they and the seeds pass through the badgers' digestive system very quickly, and the analysis confirmed that the seeds remained undamaged during their passage.

ABOVE With an estimated age of 5,000 years, the Defynnog yew (*Taxus baccata*) in St Cynog's churchyard in Wales is thought to be the oldest living tree in the United Kingdom and the oldest yew in Europe.

A Pocket Full of Rye

English crime writer Agatha Christie (1890–1976) qualified as an apothecary's assistant in 1917 and worked as a pharmaceutical dispenser in both world wars. She was thus familiar with the drugs and poisons of the early twentieth century, and subsequently wove them into the plots of some of her 66 detective novels. *A Pocket Full of Rye* (1953) tells of the events at Yewtree Lodge, where the wealthy Rex Fortescue, his latest wife and the maid (who had been poisoned by the amateur sleuth Miss Marple) are all murdered, the first death being due to taxine poisoning. In this twisting story of hidden identities and generational vengeance, the murderer needs to solve the problem of the bitter-tasting yew toxins. To mask this telltale sign, the poison is mixed in English marmalade made of Seville oranges, whose peel already imparts a slightly bitter taste to the spread.

Fact files
Details of the featured plants, including their scientific name and the author (following Kew's Medicinal Plant Names Services; see pages 15 and 219); commonly used scientific names that are no longer current follow in parentheses with the designation 'syn.' for synonym; plant family (following the Angiosperm Phylogeny Group IV; see page 219); most widely used common names; the type(s) of toxin they contain and in parentheses the most important or abundant compounds of each type; and lastly, the symptoms they cause, divided into parts of the body and ordered from least to most serious. Symptoms usually relate to poisoning in humans after a single dose or ingestion, but if they follow other circumstances or affect other animals this will be indicated.

Boxes
Used to focus on a particular poisoning case or an item of more general interest.

Arrangement

Although the end result of eating these killer plants is the same, they cause death by targeting a variety of organs and systems in the body. These targets are used to group the plants into chapters within the book, with the first pages of each chapter introducing the diverse mechanisms of action of the compounds it covers. Subsequent pages explore each type of compound, and feature the plant or plants that cause the most serious or numerous poisonings. Other plants are also included if they have an important place in humanity's relationship with these harmful compounds. Chapter 10 is slightly different, as it looks at some of the numerous plants whose toxic compounds are used by humans as medicines or insecticides.

Some plant families include a large number of poisonous plants, and in an interesting twist many of them are also important sources of plants that are commonly eaten. These families are introduced on coloured feature pages, interspersed throughout the book. They provide an overview of the family, named according to a recent international classification (see page 219), and link together the plants from that family that are covered elsewhere in the book.

CHAPTER 1

WHY SOME PLANTS ARE TOXIC

Plants cannot run away from herbivores that feed on them, or fungi, bacteria and other microorganisms that attack them, and so they need other ways to protect themselves and fight back. One such strategy is by chemical means, producing poisonous and noxious compounds that deter feeding and infection. This chapter will explore what a plant is and explain how we use classification and nomenclature to describe plant diversity, why and how plants might produce toxins and why they themselves are immune to these poisons.

Plants and their diversity

Before we look at the world's most poisonous plants and the toxins they produce, we need to consider what we mean by a 'plant'. The classical division of the living world is into moving animals and immobile plants, but how does that hold up in a modern sense? With the advent of the microscope we discovered that, although they are invisible to the human eye, single-celled organisms are abundant. And by applying chemical and, more recently, genetic analyses we now realize that some sessile organisms are more similar to animals than to plants. So what defines plants and distinguishes them from animals?

What is a plant?

Many of us consider that the most prominent feature of plants is their green appearance. As discussed later in this chapter, this is a consequence of their ability to perform photosynthesis, the process by which plants utilize the energy in sunlight to convert carbon dioxide and water into sugars. However, this process, which releases oxygen as a by-product – and which is extremely important for animals like us – is not unique to plants. There are a number of bacteria, called cyanobacteria, which also perform this task. In fact, the chloroplasts, the plant organelles responsible for photosynthesis, are ancient cyanobacteria trapped inside plant cells. There are also a number of plants that do not perform photosynthesis, but live as parasites on other plants or rely on fungi to provide them

Parasitic and hemiparasitic plants

This book is about plants that kill, and focuses specifically on plants containing compounds that affect other organisms, such as fungi, grazing animals and, of course, humans. However, there are some plants that harm other plants, either through compounds they release into the surrounding area (see page 143 for an example) or by being parasitic or hemiparasitic on them. Such plants come in many different forms, ranging from the world's largest single flower, a species of rafflesia (*Rafflesia arnoldii*) pictured below, with a diameter of about 1 m (3 ft), through rather normal-looking witchweeds (*Striga* spp.), which can cause havoc in fields of cereal crops, to mistletoes (such as *Viscum* spp. and *Phoradendron* spp.), which live only on the branches of other woody plants. Of these examples, rafflesia is totally parasitic, taking all its nutrients from the host plant, while both the witchweeds and the vast majority of mistletoes do still carry out some photosynthesis and can produce some of their own nutrients.

Due to their strong dependence on the hosts, it is disadvantageous for these parasites and hemi-parasites to be true plant killers, but as some parasitic plants are occasionally used as food by humans, this presents another kind of toxicity risk. As they draw water and nutrients from their host plants, they may also sequester other compounds from the host. Consequently, if a parasitic plant is growing on a poisonous host, it often becomes toxic itself. For example, mistletoes growing on oleander (*Nerium oleander*) contain cardiac glycosides, and if the host is a corkwood tree (*Duboisia* sp.), they sequester nicotine. This absorption of host toxins explains why Native American people usually eat mistletoe berries only if the plant is growing on a known non-poisonous host.

ABOVE **Rainforests, such as this one in Queensland, Australia, are regions of high biodiversity, being rich in the number of species of plant and other organisms that are found in them.**

with nutrients. Though parasitic plants are only rarely killers, albeit of their host plant, they can cause severe damage, including to crop plants (see box).

Plant cells differ from animal cells in that they have a cell wall and not just a membrane (see page 30 for a comparison). However, the presence of a cell wall is not unique to plants; bacteria and fungi have them as well. What is unique about plant cell walls when compared with all other living things, and is in fact the one character found in all 'plants', is that they are constructed from carbohydrates, the most common of these being cellulose. Bacterial cell walls, on the other hand, contain proteins. Fungi, once considered to be plants, use chitin, which is the same compound that makes up the exoskeletons of insects and crustaceans, supporting the notion that fungi are more closely related to animals. Some plants further strengthen their cell walls by using lignin, the compound responsible for woodiness, or suberin, which builds up cork tissues.

Plant diversity

Within the plant kingdom, the Plantae, there is a great diversity of form, which you will appreciate if you look around a garden or park, or go for a walk in the countryside. This diversity is recognized as a number of major groups within the kingdom, some of which will be familiar. The most abundant and variable group is the flowering plants, the angiosperms. Most of the plants in this book belong to this group, which has evolved the most complex range of chemicals as protection. Other groups, such as cycads, ferns and conifers, have far fewer representatives of significance as poisons, and the remaining groups, including mosses, liverworts, hornworts and lycopods, do not get a mention in this tribute to the world's killer plants.

Plants come in all sizes, from single-celled algae to trees more than 100 m (300 ft) tall. But even the multicellular plants rely heavily on passive physical processes to transport water and nutrients, while animals use muscles and a circulatory system to do the same. Plants use concentration gradients to create an osmotic pressure, drawing water up through their roots, and evaporation from the leaves, called a 'transpirational pull', enables transportation to the furthest reaches of the organism (see also 'Root to leaf', pages 18–19).

Classification and nomenclature

As humans, we instinctively name and classify the things and concepts that are important to us, and in doing so bring order to, and are able to communicate about, an otherwise chaotic world. Plants were important to the survival of our ancestors – some were sources of food, while others needed to be avoided as they were harmful. We can imagine that different groups of people had their own systems of naming and classifying, initially quite simple but becoming more sophisticated as language evolved. The ideas underlying the classifications that were developed can still be seen in the common names we use for plants.

Many plant names are descriptive to assist with identification, incorporating characteristics such as colour, size or texture, when the plant flowers, or whether it bears fruit. The use of particular terms could denote that a plant is harmful, such as nightshade, and might even indicate which animal the plant was thought to kill or harm, such as wolfsbane and dog's mercury.

THE FATHER OF TAXONOMY

With the advent of the Renaissance, the established language of scholars was Latin. Thus, in 1735 when Swedish botanist Carl Linnaeus (1707–1778) published his *Systema Naturae*, which laid out classification schemes for plants, animals and minerals, he did so in Latin. In fact, until 2012 descriptions for new species of plant had to be written in Latin in order for their publication to be valid; now, both Latin and English are permitted.

Within the plant kingdom, Linnaeus devised a 'sexual system' of 24 classes based on the number and characteristics of stamens, or 'husbands', with each class divided into orders determined by the number and position of pistils, or 'wives'. Linnaeus further divided each group into genera, and the genera into species based on morphological similarity. This is an example of a 'hierarchical classification'.

Linnaeus considered his classification to be artificial and believed that with further study it would be improved or replaced. Subsequent botanists undertook this work, for example the French botanist Antoine Laurent de Jussieu (1748–1836), who, with the publication of *Genera Plantarum* in 1789, introduced families as a level between genus and order. We still recognize several of his 100-plus families today.

LEFT **The unassuming twinflower (*Linnaea borealis*) from the northern hemisphere was Linnaeus's favourite plant. It was named by his teacher and published by Linnaeus in 1753 (see box).**

FAR LEFT **Carl Linnaeus (Carl von Linné) wearing twinflower (*Linnaea borealis*) on his jacket lapel.**

Botanical binomials

Scientific names are 'binomial', consisting of the genus name and a specific epithet to distinguish between species. Although others had used binomials, Linnaeus was the first to adopt them consistently. His 1,200-page book *Species Plantarum*, published in 1753, is now used as the starting point for the scientific names of species and their descriptions.

Botanists still use binomials to name plant species because they have proved to be useful. In contrast to common names, binomials are written in a single language, that is 'Latin form', even though the 'root' of the name may be from any language, and are recognized and used internationally. Since 1753, more than 900,000 species names have been published for an estimated 370,000 species of plant. There are more names than plants because a botanist publishing a name for a plant might have been unaware that it had already been named, and also because views on what constitutes a 'species' have differed. It is the work of today's botanical taxonomists to describe and delimit plants using the specimens and techniques available to them, and to determine the correct binomial for each following the rules of botanical nomenclature. Usually, the name (the specific epithet, if not the binomial) that was published first for a plant has priority and is called the 'accepted' name, with any later published name considered to be a 'synonym'.

Because there are instances of the same binomial inadvertently being published by different botanists for different species, the author who published a name is included after the binomial in scientific publications to make it clear which use of the binomial (which 'homonym') is intended. The author names are often abbreviated, as in the use of 'L.' for Linnaeus. This book includes the authors of scientific plant names only in the fact file for a particular species. Following convention, the binomial portion is italicized but the author is not, e.g. *Aconitum ferox* Wall. ex Ser.

Example classification of a species (levels above family can vary depending on the authority followed):

Kingdom: Plantae

Division: Magnoliophyta

Class: Magnoliopsida

Order: Solanales

Family: Solanaceae

Genus: *Atropa*

Species: *Atropa bella-donna* L.

INCREASINGLY NATURAL

Artificial classifications are usually based on a few characters, so that species, which are grouped by those classifications, may share few other characteristics. Later taxonomists worked towards more natural classifications by increasing the number of characters used to construct them, including phytochemical, microanatomical and chromosomal information. Such classifications are 'phenetic', as they are based on overall similarities and differences between the species as they exist today.

Current classifications usually aim to reflect the evolutionary history and relationships of plants, and are termed 'phylogenetic' classifications. International initiatives, such as the Angiosperm Phylogeny Group (see page 219), have used comparative gene sequence information to collate a classification for all orders and families of seed plants, both flowering plants (angiosperms) and conifers and their relatives (gymnosperms).

Evolution

Most of us have a notion of what evolution is, but many of us know little beyond the misconception that humankind is derived from apes or the slogan 'survival of the fittest' and the term 'natural selection'. What evolution aims to explain is how the diversity of life, with all its species, came about.

LEFT **In the peloric form of the yellow toadflax (*Linaria vulgaris*), the gene that usually ensures the flower becomes bilaterally symmetrical is shut off, causing it to develop a radial symmetry instead.**

LEFT **The normal form of the yellow toadflax (*Linaria vulgaris*) has bilaterally symmetrical flowers, with one plane of symmetry. This erect perennial is native to Europe and parts of temperate Asia.**

Challenging concepts

The principles introduced by Linnaeus during the eighteenth century to order and describe the diversity of life – principles we still use today – had a theistic dimension. It was thought that organisms had been created to perform certain roles and to fulfil specific purposes. When Linnaeus was presented with a specimen of the yellow toadflax (*Linaria vulgaris*) that had deformed flowers, he struggled to explain its place within the plant kingdom and named it *Peloria*, based on a Greek word for monster. Linnaeus considered species to be unchanging and stable, but this form of the yellow toadflax indicated that species could change. The botanist settled on the thought that this strange flower was the result of hybridization between yellow toadflax and another, as yet undiscovered, species.

On the Origin of Species

It was not until the publication of the book *On the Origin of Species* in November 1859 that naturalist Charles Darwin (1809–1882) firmly launched the scientific theory of evolution, even if he actually did not use this specific word until the sixth edition. Here, the thought that organisms struggle to survive, and that any individual variation in traits making survival and reproduction more probable will be favoured and fixed in future generations, was laid out. This also means that closely related species share a common ancestor, which is the case for humans, apes and other primates. Though the actual mechanisms underlying the inherited traits were unknown at the time, the theory elegantly explained observations from embryology, animal husbandry and biogeography. The concept of natural selection had actually been propounded in a reading at the Linnean Society of London the previous year, where Darwin's work was presented together with that of Alfred Russel Wallace (1823–1913), who is considered to be the father of biogeography and often accredited as co-discoverer of the theory of evolution.

INHERITANCE

Independently of the theory of evolution, but concurrently, an Augustinian friar in the Austro-Hungarian Empire was studying peas. His name was Gregor Mendel (1822–1884), and it was his experiments that eventually led to a greater understanding of how traits are inherited. In his work, Mendel crossed peas with different flower colours and positions, and different seed shapes and colours. By evaluating, over several generations, the number of offspring sharing the traits of their parents, he came to the conclusion that some invisible factor determined these traits in the progeny. Mendel also concluded that these inherited factors came in pairs, one from the maternal parent and one from the paternal, and that there were two sorts, one that would determine the trait if passed on from only one parent, and one that would affect the trait only if derived from both parents. He called traits needing to be inherited from both parents 'recessive', and those from only one parent 'dominant'. Sadly for Mendel, his discoveries would not be recognized until after his death. However, we now call his invisible factors genes and, after the discovery of the structure of deoxyribonucleic acid (DNA) in 1953, studying them has influenced almost every single field of biological research.

ABOVE Cacti and succulents have evolved a number of mechanisms that enable them to survive in arid conditions, such as water-storing stems and thick cuticles to reduce water loss through evaporation.

RIGHT Gregor Mendel discovered the basic principles of heredity in his monastery garden.

BELOW Diagram showing inheritance of flower colour in peas. Mendel used peas in his experiments due to the large number of different forms, and the short time it takes to raise a new generation of plants.

THE SUCCESS OF EVOLUTION

The selection of genetic traits inherited by future generations, which may cause organisms to evolve into other species in the long run, is influenced by a number of factors. As plants are sedentary organisms they are dependent on the local environment, and these ecological factors can influence the selection of particular traits. For example, the water-storing stems of cacti and succulents (see pages 116–117) have evolved to cope with similar dry climates.

Being rooted to the ground also means that plants can't run away from pests and hungry animals. Evolution has, however, provided several ways for plants to defend themselves. Woodiness, thorns and stinging hairs give protection against most grazing animals, but fungi and insect pests are not equally deterred. Another plant strategy, which will be the focus in the rest of this book, is to produce compounds that are toxic to herbivores and infectious organisms. However, evolution is at work on all organisms, not just plants, so animals can evolve mechanisms that enable them to forage on poisonous plants (see pages 23 and 35) or even take advantage of plant poisons to protect themselves from predators (see page 43).

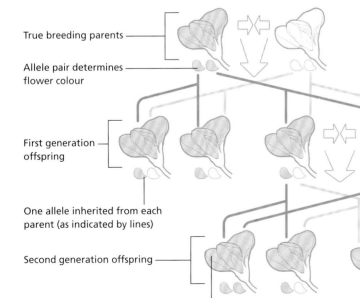

Root to leaf

The first organs to emerge from a seed are the root, stem and leaves. Here the functions of these vital parts of the plant will be outlined and what makes them potentially poisonous explored.

Below the ground

Roots anchor plants to the ground or, in the case of epiphytes, suitable crevices on tree branches, and are responsible for taking up water and minerals such as nitrates. They come in several forms, and while some plants form a root network near the soil surface, absorbing water and minerals before they percolate through the soil layers, other plants might develop tap roots that grow down into the soil and can draw on deeper supplies.

Roots, particularly tap roots such as carrots, also often function as storage organs. They build up reserves of carbohydrates such as starch and other nutrients that can be kept safe when the plant is dormant, for example, over the winter in temperate regions of the world. This is then broken down and transported to the rest of the plant during the active growing season. New storage organs then develop, growing in size as they accumulate starch in preparation for the next dormant period. There are several other forms of underground storage organs, most of which would not be categorized as 'roots' by a botanist. For example, tubers, such as sweet potatoes, are considered modified roots; bulbs, such as onions, are formed from scale-like leaves; and corms, rhizomes and stolons, including taro, ginger and potatoes, respectively, are underground stems.

As absorption of water and storage of nutrients are essential for the survival of a plant, the underground organs often have physical barriers, such as calcium oxalate crystals, and deterrent compounds synthesized or accumulated in their tissues, to protect against herbivory and attack by pathogens, including bacteria and fungi, or by larger organisms such as eelworms.

Stems

Plants usually have some sort of stem, which can be slender and more or less pliable in herbaceous plants and the new twigs of trees and shrubs, or rigid in woody stems and branches, and the trunks of trees. A function of many stems is to hold the leaves above the ground and positioned to capture sunlight. Their other main function is the movement of water and food around the plant. This is made possible by a vascular system of specialized cells that form continuous tubes: the xylem transports water and dissolved minerals from the roots to the aerial parts of the plant, while the phloem transports sugars around the plant to all parts.

As the phloem tissue transports large amounts of sugar, it is subject to frequent attacks by insects, fungi and bacteria, which feed on the nutritious solution it contains. To protect the vascular system, it is supported by woody fibres and, in some plants, by ducts running along the length of the stems. When damaged, these ducts exude a sticky resin or irritating latex that contains compounds with properties that can reduce infection and discourage herbivores and insects from feeding. Examples of very potent noxious exudates include those in plants such as the spurges (*Euphorbia* spp.; see pages 114–119) and the opium poppy (*Papaver somniferum*; see pages 200–201).

LEFT **Parsnip (*Pastinaca sativa*; see page 129) tap roots store carbohydrates and enable the plant to survive over winter. Cultivated forms have been bred with bigger tap roots and a milder flavour.**

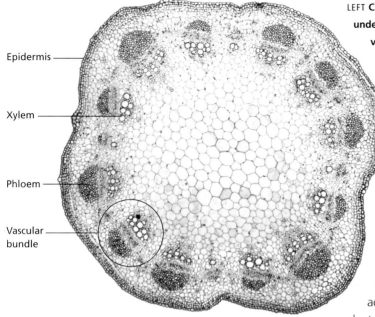

LEFT **Cross section of a sunflower (*Helianthus annuus*) stem viewed under a microscope. The central pith is surrounded by a ring of vascular bundles, formed from xylem and phloem cells.**

LEAVES

The leaves of a plant are usually its most obvious part, especially during the growing season in temperate regions. They are also usually the organs in which photosynthesis occurs (see pages 24–25). In addition to vessels for water and nutrient transportation, the leaves contain a spongy tissue in which carbon dioxide and oxygen gas exchange can occur through openings called stomata on the leaf surface. Leaves can arguably be considered the plant's most important organs, and evolution has come up with several strategies to protect them from herbivory, infection by microorganisms and ecological threats such as drought.

In hot, dry climates, leaves might be protected by thick waxy layers or hairs to prevent water evaporation, or be disposed of completely, as in most cacti, where the stem has taken over as the location for photosynthesis. And some plants have developed hairs that contain strongly irritating compounds, or even function as tiny syringes, injecting noxious substances to deter herbivores (see Chapter 6). In some leaves, insoluble crystals, fibres and compounds such as tannins, which hinder herbivores and pests from actually digesting the leaves, are accumulated. Yet other plant leaves are poisonous due to the presence of particular compounds (see pages 26–27).

The level of toxicity in leaves may vary under a number of conditions. Young leaves are often more toxic than older ones, as observed in some plants whose leaves show damage from herbivores only in the autumn, once the leaves have fulfilled their function and are about to be shed. Toxins can also be increased by specific triggers. Attack by pathogens or herbivores can elicit the production of compounds that are toxic to the specific organism. Such compounds, called 'phytoalexins', include the furanocoumarins (see page 128) that are produced in celery (*Apium graveolens*) when it is infected by fungi. Ecological factors like drought may also induce the production of poisons, as exemplified by the bitter cucumbers (see pages 150–151).

BELOW **Diagrammatical cross section through a leaf, showing the specialization and arrangement of cells to maximize the capture of sunlight and reduce water loss during the exchange of gases.**

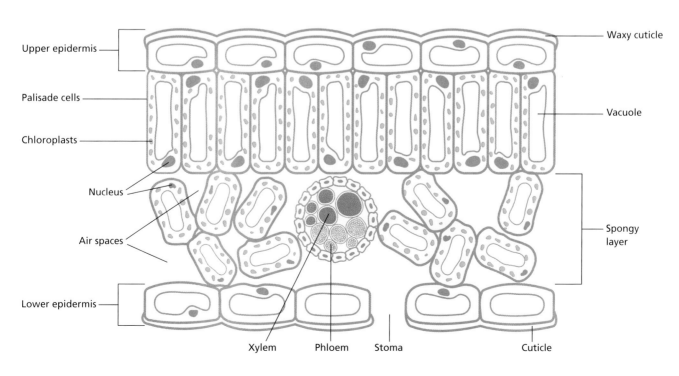

Flowers

Most killer plants are what we class as flowering plants (see pages 12–13), so although some plants, such as ferns, use other structures and mechanisms to reproduce, the plants that we are interested in rely on flowers for this vital process.

Reproduction

The male parts of the flower are called stamens. A flower usually has more than one stamen, each consisting of a filament with an anther at the tip. The anther produces pollen, the male gametes. The female parts of the flower are called carpels, with one or several fused together to form a pistil. Each flower usually has only one pistil. The pistil is formed from a stigma at the tip, to which the pollen will stick, an ovary at the base, and a tube called a style joining the two. The ovary contains the ovules, or eggs, the female gametes.

Unlike animals, plants can't move in search of a suitable mate. Some plants, including the grasses and many trees, rely on wind to blow their pollen onto the stigmas of other flowers that may be some distance away. This transfer is facilitated by production of large quantities of light pollen grains by heads of flowers, such as catkins, which are held aloft, and by large feathery stigmas, which sweep the air as the wind blows. Such flowers are usually small and lack showy parts so as not to obstruct the free movement of pollen.

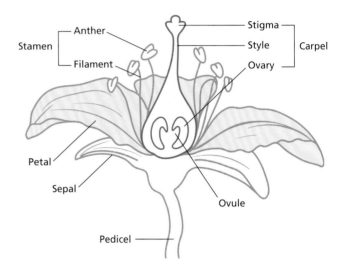

ABOVE **Diagram of a flower with male (stamen) and female (carpel) parts enclosed within a ring of petals. Differences in the number and arrangement of floral parts are used to identify plants.**

Pollinator attraction

In plants that depend on insects or other animals to transfer pollen from one flower to another, variations in form are greater. Flowers have evolved with structures and characteristics that favour particular pollinators, and pollinators have co-evolved to visit a limited number of flower types, sometimes favouring only one particular species. Flowers attract pollinators using a combination of scent, colour and food reward tailored to the particular pollinator.

RIGHT **Flowers attract pollinators, such as this fritillary butterfly (*Argynnis* sp.), which pollinates the flowers of cowbane (*Cicuta virosa*) while drinking their floral nectar.**

ABOVE **Monkshood (*Aconitum napellus*) flowers have poisonous sepals to discourage nectar robbing and encourage pollination by the long-tongued garden bumblebee (*Bombus hortorum*).**

Scent plays a major part in the attraction and selection of effective pollinators. Fruity and flowery odours are usually attractive to butterflies, moths and bees, while musty and rotten smells attract flies and beetles.

Colour is usually provided by showy petals, but in some flowers other parts have this function: the sepals, the outer parts that protect the flower bud, are usually green, but in clematis (*Clematis* spp.) and aconite (*Aconitum* spp.) they are coloured; some flowers do not have separate petals and sepals but have tepals instead, such as in the tulips (*Tulipa* spp.); and bracts, which are modified leaves, can be the most obvious coloured part, as seen in poinsettia (*Euphorbia pulcherrima*). Due to differences in the colour vision of pollinators, some will be attracted to certain colours. For example, birds usually choose red or orange flowers, while bees see better in the blue to ultraviolet end of the colour spectrum. Many moth- or bat-pollinated flowers are white, as this colour is more visible at low light levels.

POLLINATOR REWARD

Pollinators are rewarded for visiting a flower, usually with nectar, a sweet, sugar-rich substance. In some flowers the nectar is easily accessed by insects with short mouthparts, such as ants and flies, while in others it can be more difficult to reach and requires a longer tongue, or for the insect to crawl into the flower. Whatever the case, the parts of the flower will be structured and arranged so that while accessing the nectar the pollinator will pick up pollen. It will then transfer this pollen to the stigma of the next flower of the same species it visits. By limiting the type of pollinator that visits, the flower benefits from an increased chance of successful pollination, and the pollinator does not waste time and energy visiting a flower that will not reward it.

DISCOURAGING NECTAR ROBBING

The production of flowers, seeds and fruit requires a significant allocation of the resources produced by the plant's leaves, but it is necessary for reproduction. So, in addition to increasing the chances of successful pollination as described, plants also protect their flowers in a variety of chemical ways. Some inventive insects and animals try to bypass the structures of the flower that enable only pollinators to access the nectar. They usually do this by eating through the petals or other protective parts of the flower from the outside. To discourage this, some plants have toxic petals, as are found in angel's trumpets (*Brugmansia* spp.; see page 83) and aconites (*Aconitum* spp.; see pages 48–49), while others even produce toxic nectar, such as rhododendrons and azaleas (*Rhododendron* spp.; see pages 78–79 to find out why they are still successfully pollinated).

Fruit and seeds

Successful pollination results in the fertilization of the ovules by pollen from a plant of the same species. The ovules then develop into seeds that remain protected within the carpel, which grows to form the fruit. However, just as plants rely on external factors for pollination, so they are also unable to influence actively the kind of start their progeny get in life. Some fruits and seeds are passively dispersed by wind or water, others get stuck in the fur of animals, and some even need to be ingested and will not sprout unless they pass through the digestive tract of an animal. There are even plants with fruits that are explosive or use spring-like structures to fling their seeds into hopefully fertile ground. With such different dispersal mechanisms, there is understandably a large diversity of fruits.

Worth protecting

The toxins in fruits and seeds are there to protect the 'next generation', and the amount and type of toxin might vary a great deal between different stages of ripening. As seeds themselves are an attractive source of food for animals, containing the nutritious plant embryo and usually also an endosperm as a source of food for the seedling, both developing and mature seeds can contain compounds that deter predators (see, for example, the mustard bomb, pages 120–121) or protect against fungal infection and other pests. In addition, mature seeds are usually protected by hard seed coats, which also prevent the embryo from drying out before conditions are suitable for its germination.

ABOVE **The dry capsules of opium poppy (*Papaver somniferum*) release numerous small seeds through a ring of apertures that open when the seeds are ripe. The seeds do not contain enough alkaloids to pose any risk.**

LEFT **Jimsonweed (*Datura stramonium*) capsules are protected by their covering of pointed spines. When ripe, they split open from the top into four sections to expose the seeds that are held within.**

Fleshy fruits and dry fruits

Fruits that are dispersed by animals after being eaten need to regulate their toxicity during their different developmental stages. There is a fine balance between being poisonous enough to keep off fungal infections and pests, while remaining tasty and desirable for the animals responsible for dispersal of the seeds. So, as the seeds inside the fruit ripen and become viable, the fruit also undergoes changes.

Fruit can be either fleshy or dry, although humans often consider only the former to be 'fruit'. During the ripening of a fleshy fruit, the carpel thickens and at least the outer layer often turns from green to more showy colours: yellow, orange, red or even blue. Antifeedant compounds such as astringent tannins, which make it harder for predators to break down and absorb nutrients, and bitter principles, making the fruit unpalatable, decrease in concentration. Correspondingly, cell walls soften, acidity changes and the amount of sugar increases to make the fruit more palatable. In many cases, the seeds may still be dangerous to eat even if the fruit becomes less poisonous over time, as can be seen in peaches and bitter almonds (see pages 182–183). In some cases, plants produce fleshy parts that are not formally fruits but nevertheless help to disperse toxic seeds (see pages 50–51 and 72).

Dry fruits also change during the development of the seeds. The forms of dry fruits are very diverse, and include nuts, samaras (winged nuts), pods and capsules. Instead of the carpel becoming more palatable as in the fleshy fruits, here it becomes more fibrous, dries out and develops a hard protective sheath for the seed or seeds. Some dry fruits are dispersed by animals – residents of the northern hemisphere, for example, will be familiar with nut-foraging squirrels – but it is more common that they rely on passive dispersal. They may produce wings or a feathery structure to enable wind dispersal, and capsules and pods break open or develop holes so the seeds can be dispersed when the plant is rattled by animals or moves in the wind. Such fruits are seldom poisonous, but the seeds might still contain potent toxins, mainly to protect them until germination.

The human risk of poisonous fruits and seeds

Even when plants rely on animal vectors to disperse their seeds, there are several poisonous fruits in the world. How can this be? The berry of the deadly nightshade (*Atropa bella-donna*; see pages 80–81) is very poisonous to humans even after cooking. However, the plant's toxins have less effect on birds, which are effective seed dispersers. Interestingly, in what is an example of a herbivore evolving a defence against poisoning, rabbits are one of the few mammals that are unharmed by this plant. Some rabbits produce the enzyme atropinesterase, which breaks down the toxin atropine found in deadly nightshade before it can exert its effect. But even if it is safe for rabbits to eat the plant, it might not be safe for predators to eat such rabbits soon afterwards – there are reports of humans being poisoned after eating rabbits that have recently grazed on atropine-containing plants (see also page 101 for another example of hemlock poisoning).

ABOVE **Ripe fruit of deadly nightshade (*Atropa bella-donna*), cut longitudinally to show its purple juice and the numerous seeds within. There is a star-shaped ring of five large leafy sepals at the base of the fruit.**

LEFT **The flesh of the peach (*Prunus persica*) fruit is juicy and edible when ripe, attracting animals that can spread the seed, but surrounds a hard 'stone' that encloses the toxic kernel.**

Photosynthesis and metabolic pathways

As mentioned at the beginning of this chapter, plants are usually green due to the presence of chloroplasts, cell organelles that perform photosynthesis. This process captures the energy influx from sunlight and transforms it into chemical energy by producing sugars and oxygen from water and carbon dioxide. Sugars are one of the primary sources of energy for most organisms not capable of photosynthesis. An animal's ability to utilize sugars for energy production requires oxygen and produces carbon dioxide, thus in some way repaying plants in the process.

Photosynthesis

Photosynthesis can be viewed as a two-step process. In the first part, chlorophyll, the pigment that gives chloroplasts their green colour, absorbs light, which in turn releases an electron that is ultimately recycled by splitting two water molecules into oxygen and protons. The protons are used to produce the universal energy carrier compound adenosine triphosphate (ATP), which is used in almost all energy-dependent cellular functions, including the second part of photosynthesis. This second part, which is often called the 'dark reactions' as it is not light-dependent, is responsible for fixation of carbon dioxide. The first of these reactions is performed by what is probably the world's most common enzyme: ribulose-1,5-bisphosphate carboxylase/oxygenase, or RuBisCO for short. This enzyme fuses carbon dioxide, taken up through the stomata of the leaves, with ribulose, a sugar compound, and produces two molecules of phosphoglycerate. These are then used either to re-form ribulose, enabling a new carbon dioxide-fixation reaction, or to produce the sugar glucose, which can be incorporated into cellulose and starch, and subsequently used for energy by other organisms. To produce one molecule of glucose, RuBisCO needs to perform six reactions between carbon dioxide and ribulose. RuBisCO is an enzyme that will function less efficiently if a large amount of oxygen or high temperatures are present. Some plants have thus developed modifications of their anatomy and biochemistry to enhance the efficiency of the process. Plants living under very dry, warm conditions need to keep their stomata closed during the day in order not to lose excessive amounts of water. However, this will also prevent them from taking up atmospheric carbon dioxide. Many such plants, for example all cacti, including the peyote cactus (see pages 86–87), have

LEFT **Trees in a forest grow up towards the sun so that the chlorophyll in their leaves can capture the light and use it to drive the process of photosynthesis. Little light reaches the forest floor.**

therefore developed a process whereby during the cooler night they open their stomata to allow carbon dioxide to diffuse inwards with a minimum loss of water. They then bind the carbon dioxide in the form of an acid, which is used as the source of the gas for RuBisCO during the day.

Metabolic pathways

In addition to providing oxygen and sugar, chloroplasts also synthesize the majority of amino acids necessary for a plant to make proteins. This is almost of equal importance, as plants produce essential amino acids that animals, being unable to make them for themselves, need to ingest through their diet directly or by eating animals that are herbivores. Among these compounds, the three aromatic amino acids – phenylalanine, tyrosine and tryptophan – are especially important. They are, for example, necessary for humans to produce signalling compounds regulating heart rate and blood pressure, as well as brain functions relating to movement, mood and sleep. In plants, they are frequently used to produce toxins, such as the well-known strychnine (see pages 66–67), some types of curare (see pages 94–97) and morphine (see pages 200–201). There are several other differences between the metabolic pathways in animals and plants besides the production of sugars and all the amino acids present in animal proteins. One that is relevant to plant toxicity is the presence of isoprenoid compounds, built up of isoprene units of five carbon atoms. Although these include steroids that can be produced by animals, plants have developed a rich diversity of these substances, the terpenes. Plants also produce a range of compounds called acetogenins and polyketides, which are derived from acetic acid precursors. In animals, this metabolic pathway is primarily used to produce fatty acids, but in plants the compounds may also be turned into a wide array of substances with diverse functions. Further discussion on these and other poisonous plant compounds follows on the next page.

ABOVE **Artificially coloured transmission electron micrograph (TEM) of leaf cells, showing that each has a single nucleus (brown) and several chloroplasts (green) to perform photosynthesis.**

Compartmentalization

Compounds produced by a plant that deter or harm herbivores or have antimicrobial properties can also be harmful to the plant itself, with the potential to disrupt essential cell processes. This is avoided by 'compartmentalizing' these toxins, for example storing them in the cell vacuole. The vacuole is the largest organelle in a plant cell and can occupy 30–90 per cent of the cell volume. When full of water, vacuoles contribute to the rigidity of a plant, and in addition to toxins they store nutrients and other compounds such as pigments. Of course, when herbivores damage plant cells, these toxins are released from the vacuoles and cause harmful effects.

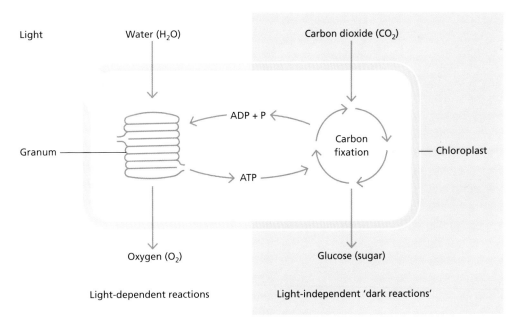

RIGHT **Simplified diagram of the process of photosynthesis within a chloroplast. During the first step, energy from light is used to produce adenosine triphosphate (ATP) and oxygen. In the second step, which does not require light, ATP releases energy to enable the conversion, or fixation, of carbon dioxide into the sugar glucose in a process known as the Calvin cycle. It is necessary for almost all life on Earth.**

Small molecular compounds

Just as we divide plants into taxonomic groups and species, so we also divide chemical compounds into different groups and substances. But in contrast to plant classification and evolution (see pages 14–17), there is no evidence that chemical groups are hierarchical or evolutionary, in the sense that we cannot trace all compounds back to a single common ancestral substance.

A common distinction is often made between 'primary' compounds, responsible for basic functions necessary for cell and organism survival (for example, sugars, proteins and fats), and 'secondary' compounds, which might give a species an evolutionary advantage but are not essential. Rather than using this functional division, we can instead group substances based on their chemical properties or how organisms synthesize them.

As previously indicated, there are differences between how animals and plants go about making chemical compounds. Looking specifically at protective compounds, animal venoms are often based on proteins, which are large compounds built up of amino acids. In contrast, plants use amino acids and other compounds as precursors, or starting material, to make smaller molecules for their poisons. Almost all examples of substances responsible for plant toxicity can therefore be categorized as small molecular compounds (although see pages 146–149 for toxic plant proteins). Here we give a very short introduction to the different groups of small molecular compounds most relevant to the poisonous plants in this book.

POLYKETIDES AND ACETOGENINS

Animals use acetic acid-derived precursors to produce fats, but plants can use the same precursors to make a whole range of other compounds. Depending on how much of the original acetic acid molecules are retained, the resulting compounds are called polyketides or acetogenins. These include the practically non-toxic flavonoids in flower pigments and the astringent tannins that can cause stomach upsets. Some plants produce even more potent acetogenins that have severe laxative effects (see pages 144–145). Another example of toxic acetogenins is the pro-allergenic compounds in the cashew family (see pages 130–131).

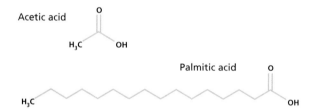

ABOVE **Acetic acid is important in several cellular processes and is used as a starting point in the production of plant compounds called acetogenins or polyketides. One example of an acetogenin is palmitic acid, the main fatty acid in palm oil from which it was first isolated.**

LEFT **Bottles of edible palm oil and the oil palm fruit from which it is extracted. Most palm oil comes from the African oil palm (*Elaeis guineensis*).**

Gallic acid

Catechin

ABOVE Tannins are not a single compound, but are built up of glycosides from derivatives of gallic acid and catechins.

BELOW Isoprene is the common unit in terpene compounds, of which rubber consists of several hundred coupled units.

Isoprene

Rubber

TERPENES

Terpenes, named after turpentine, are made up of isoprene units consisting of five carbon atoms. They can be divided into different types based on the number of included units, although confusingly they are named according to how many pairs of isoprene units they contain: monoterpenes contain two units, diterpenes four units and so forth until you reach the polyterpenes, with 'a lot' of units. These polyterpene compounds are usually poorly soluble or insoluble in water, as might be most evident for natural rubber, which consists of enormous molecules made of several hundred isoprene units. Smaller terpenes are volatile and make up several floral scents, and though they can have very good antibacterial activity their concentrations in plants are usually too low to be problematic for humans. However, slightly larger terpenes include compounds that can severely harm or even kill by causing seizures (see pages 70–73), as well as plant steroids, of which a particular sub-group can cause deadly heart arrhythmias (see pages 54–61).

ALKALOIDS

This group is probably the most important for any aspiring plant toxicologist. The compounds in this group do not necessarily derive from the same type of precursor, but they all include one or more nitrogen atoms and have a limited distribution among organisms. The typical alkaloids are those derived from amino acids and include, among others, compounds that can cause seizures or paralysis (see pages 66–69 and 94–97), evoke hallucinations (see pages 80–83 and 86–87), make you vomit (see pages 136–137) or

RIGHT The drug *Radix Polygalae Tenuifoliae*, made from dried roots of *Polygala tenuifolia*, contains the saponin senegin III.

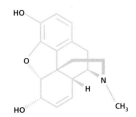

Morphine

LEFT Morphine from the opium poppy (*Papaver somniferum*) was among the first alkaloids isolated in pure form.

interrupt your cell division (see pages 152–153). But plants can also produce alkaloids by introducing nitrogen into other compounds such as terpenes. When this is done to what is usually considered a group of plant hormones, toxins causing heart arrhythmias are produced (see pages 48–49). And in the potato family (Solanaceae), incorporation of nitrogen into steroids produces alkaloids that cause severe gastrointestinal problems (see pages 140–141).

SAPONINS

These compounds share with soap the ability to foam in contact with water, and just like soap they have one part of the molecule that is easily dissolved in water and one that prefers to be dissolved in fat. They are formed by the coupling of sugars to a parent compound that has poor solubility in water. For plants, this is a common method of moving compounds around in the organism and is used for many types of parent substance. Those universally recognized as saponins usually have terpene parent compounds, and their roles in the plant are still somewhat unclear. They may be antifeedant compounds due to their bitterness, but some actually increase a herbivore's absorption of nutrients. Besides causing stomach problems, some of these compounds have been used for fishing due to their ability to kill fish by affecting the function of their gills.

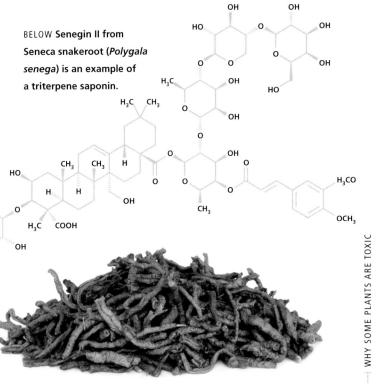

BELOW Senegin II from Seneca snakeroot (*Polygala senega*) is an example of a triterpene saponin.

Senegin II

CHAPTER 2

TARGETS IN THE BODY

There are many ways in which a plant compound can interact with another organism. In this chapter, we will give an overview of how the human organism functions and its array of targets for plant poisons. The normal functions of the different organs and physiological systems and their weaknesses can explain the potential mechanisms of toxins. The chapter will also point out certain differences in the effects of toxins on animal species and include examples, such as where insects take advantage of poisonous plants, using their compounds as a protection and deterrent against predators.

Cells and organisms

All organisms share a fundamental set of biochemical processes whether they are single-celled or multicellular, although the latter group are usually considered more advanced. In multicellular organisms, cells can be specialized and build up particular tissues, which in turn may form different organs of the body.

Building a cell

Every cell has an outer membrane, and this is accompanied by a cell wall in plants and some other organisms. The membrane basically consists of two layers of phospholipids arranged so that the outside and the inside of the cell is separated by a thin layer of fat. This functions as a semi-permeable barrier, enabling the cell to regulate the concentration of salts and other solutes it contains, and protecting it from drying out. Within each cell are several organelles, such as the mitochondrion, producing the energy of the cell, and the endoplasmic reticulum, where ribosomes produce the proteins necessary for the cell. All cells also contain the genetic material of the organism. In plants, animals and fungi, this is located in the cell nucleus.

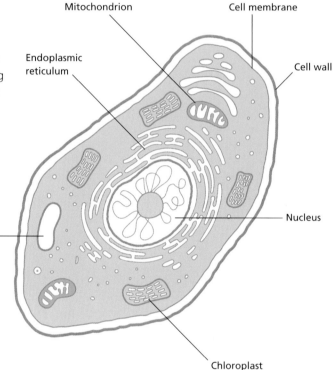

ABOVE **Anatomy of a typical plant cell. A rigid cell wall (which maintains the shape of the cell) and chloroplasts (which are the site of photosynthesis) are features found in plant cells but not in animal cells.**

ABOVE **Anatomy of a typical animal cell. Animal cells lack cell walls but share many features with plant cells, such as a nucleus, mitochondria and an endoplasmic reticulum. Vacuoles are mainly used for transportation.**

Machinery in the cell

In a basic sense, the information that defines every organism is encoded in its DNA (deoxyribonucleic acid), based on the sequence of four different compounds: adenine, thymine, cytosine and guanine. Sequences of these nucleotide bases are then transcribed and translated into proteins, which make up cells and organisms and perform their physiological processes. Any change in a DNA sequence may alter how or where proteins are produced, and if these changes affect the gametes (sex cells), they will influence the offspring and possibly, with time, the evolution of the species.

Proteins can be divided into groups by their functions, such as enzymes responsible for making or degrading other compounds, and receptors that recognize and convey signals between cells. Some proteins function as transporters, moving ions or other compounds across cell membranes or even between tissues and organs. Other proteins build up structural elements such as the bulk of muscles or the keratin in the outer protective layers of the skin and in nails and hair.

From cells to organs

Starting from a single-celled fertilized egg, the human body develops into about 30 trillion cells when fully grown (not counting up to 40 trillion cells in the form of bacteria we host, and depend upon, in our digestive canal and on our skin). In such a large organism, cells can no longer rely on absorption of nutrients from the immediate surrounding environment or on communication just with the adjoining cells. To perform these functions, specialization of cells into tissues and, subsequently, into organs is necessary. From a functional viewpoint, the primary system for animal survival can be considered the one that ensures the intake of food and absorption of its nutrients. In its simplest form, this is a tube where nutrients are absorbed as the food passes from one end to the other through the body. However, if you ingest many different types of food or if the nutrients in the food are inaccessible for direct absorption, the simple tube needs to be complemented with tissues and organs providing enzymes and mechanical ways to break down the food. In animals, this has led to the evolution of different mouthparts, the presence of one or more stomachs, differentiation of the intestines and a liver (see pages 32–35).

If the body plan of the animal involves limbs or other structures that are not in direct contact with the 'food tube', there needs to be a circulatory system using some sort of fluid to transport nutrients throughout the body. As the fluid will also absorb waste products, a filtration system that can excrete them is required. These functions are performed in humans by the blood, heart, blood vessels and kidneys (see pages 35–37). The circulatory system in vertebrates is also used to transport the oxygen absorbed in the lungs or gills out to the cells where it is needed for energy production.

This collection of organs and tissues, all of which have to interact with one another, needs a command centre to sort and interpret incoming signals and send out instructions when necessary. Such a system will also provide the signals to initiate movement, which is essential for animals as they must actively seek out their food sources. An overview of the brain and the nervous system, as well as the muscles, is given later in this chapter (see pages 38–41).

RIGHT **The toxins in poisonous plants work on different levels in the body. Drastic effects are seen when they interact with vital organs or affect whole tissues, or when they indiscriminately destroy cells.**

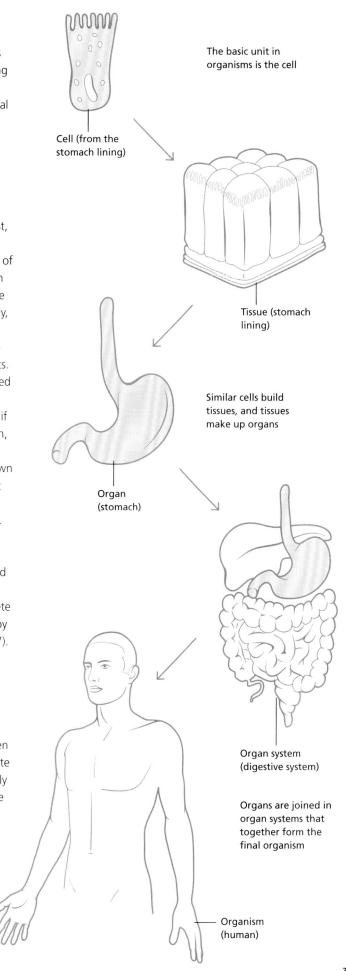

The basic unit in organisms is the cell

Cell (from the stomach lining)

Tissue (stomach lining)

Similar cells build tissues, and tissues make up organs

Organ (stomach)

Organ system (digestive system)

Organs are joined in organ systems that together form the final organism

Organism (human)

Taste and the gut

Humans, and other animals, need to eat to survive, but not everything we encounter is safe to eat. We have therefore evolved mechanisms to limit our exposure to food that contains high levels of toxins, although the number of accidental poisonings that occur suggests that these mechanisms are not foolproof.

First encounters

For humans, the sense of smell is a first test of the edibility of food, as we are repulsed by odours like those of sour milk, rotting meat or vegetables, and faeces. The repulsion evokes a feeling of unease, and if it is strong enough it can even elicit a vomiting reflex. These are clear indications that we should move away and not eat the offending object. However, even foul smells can be attractive to some animals, and certain flowers have come to rely on flies and beetles – usually those that forage in carcasses and dung heaps – as pollinators.

If the smell of a plant fails to deter herbivores, its taste or some physical attribute may limit the amount that is eaten. Plant material can contain compounds that make it bitter, astringent or sour, or even irritate the soft lining of the mouth. All these are clear signs that the plant should be avoided. If they fail to repulse feeding animals by smell or taste, plants may produce resins or latex that clog the mouthparts of smaller animals like caterpillars, effectively hindering them from destroying the entire plant.

Breaking things down

But what happens to plant material that is eaten? The function of an animal's digestive system is to break down food so that it can be used as a fuel by the body. This starts in the mouth, where food is physically broken into smaller pieces by chewing and mixed with saliva, which acts as a lubricant to ease the movement of the swallowed food along the oesophagus to the stomach. In some animals, including humans, the saliva contains an enzyme, amylase, which breaks down starch in the food into smaller sugars. Detection of these by the taste buds is an indication that the food is a good source of nutrition.

In the stomach, acidic gastric juices and a number of enzymes start the process of breaking down proteins and fats. To prevent damage to the lining of the stomach, mucus is excreted by glands to provide a protective layer.

Most of the energy in plants is bound up in cellulose and similar carbohydrates, which many animals cannot break down efficiently. Whereas humans and many other animals have a

LEFT AND BELOW **Anatomy of the tongue (left) and a taste bud (below). The surface of the tongue is covered in papillae, and it is on the surface of these that the taste buds are found. Taste buds are collections of nerve-like cells that transmit impulses to the brain.**

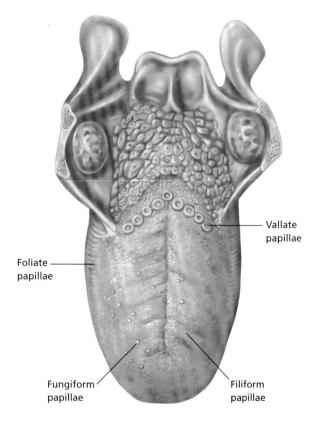

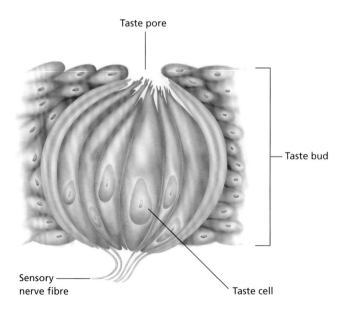

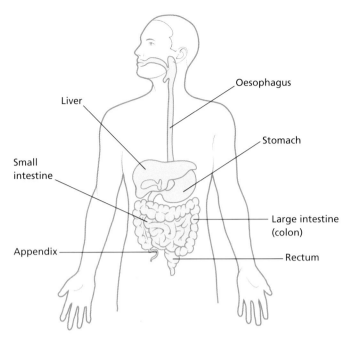

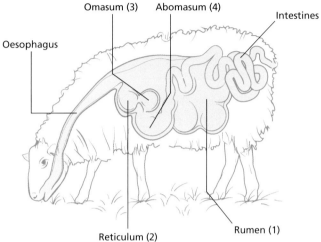

LEFT AND BELOW **The human digestive system (left) is monogastric – has a stomach with only one chamber – whereas that of a ruminant, such as a sheep (below), has a stomach with four chambers, numbered here in the order in which food passes through them.**

digestive system with a single stomach, ruminants such as cows, sheep, goats, deer, giraffe and camels have a four-chambered stomach specially adapted to digest the complex carbohydrates of their main food sources. In the first two chambers, microbes (primarily bacteria) help the animal to break down cellulose. Plants containing antibacterial substances, such as high concentrations of tannins, can therefore severely harm a ruminant's ability to take up nutrients from its feed.

Regardless of how many chambers an animal's stomach has, the food will pass into the intestines, where most nutrients will be absorbed. At the start of the intestinal canal, the half-digested, semi-liquid food is made more alkaline to protect the intestines, and additional enzymes are added to continue breaking down the food, particularly fats. Passing next to the large intestine, what remains of the food is changed to a semi-solid state by the removal of water. Any remaining protein and starch is broken down by bacteria, which make up the gut flora, and waste is finally eliminated as faeces.

Handling poisons

A toxin passing through the digestive system will undergo all the processes described above. The action of digestive enzymes can either remove the toxicity or enhance it, and the acidity at the different stages will also affect the activity of the toxin. All these factors will determine how long after the toxin is eaten any symptoms are experienced and how severe they are. The digestive system also has various mechanisms that limit exposure to a toxin once it has been detected by the body. Irritation of the stomach causes vomiting, and if the toxin is detected in the intestines, its movement through the digestive system can be speeded up and it is expelled as diarrhoea.

In order to fine-tune all its functions, the digestive system is considered to have its own nervous system, called the enteric nervous system. Although it receives many signals from the rest of the body, the system also operates independently. It is made up of nerve cells that control the muscles moving food along the digestive canal, the secretion of enzymes and the reporting of chemical and mechanical conditions.

RIGHT **The tannins in acorns and oak leaves, such as those of the sessile oak (*Quercus petraea*), can cause severe poisoning in ruminants, while monogastric animals like pigs are less sensitive.**

Liver and kidneys

As plants have evolved compounds to protect themselves, humans (and other herbivores) have evolved to cope with them. The main human organs responsible for detoxifying and getting rid of poisons are the liver and kidneys, and we explain here how these organs can be affected by these same toxins.

Liver

The liver is also considered to be part of the digestive system (see pages 32–33). Almost everything that passes through the stomach and is absorbed by the intestines subsequently passes through the liver before it is distributed to the rest of the body. The liver contributes to the health and normal functioning of the human body in a number of ways, including storing some vitamins and glucose, synthesizing cholesterol as required and glucose during starvation, and producing blood-clotting factors and bile. In addition, it detoxifies harmful substances produced during normal cell metabolism, such as ammonia, or absorbed into the blood from the digestive tract. The latter include everything from alcohol, medicines or drugs, and other chemicals, to plant toxins. During detoxification, these compounds are broken down into smaller units or transformed into water-soluble compounds, which are then more easily excreted by the body in the urine via the kidneys. If the water solubility is limited even after metabolism in the liver, the substance can be excreted in the faeces via the bile, which empties into the small intestines. However, in some instances this will prolong the duration of the poisoning as the bile is reabsorbed, returning some of the toxin to the liver in what is called enterohepatic circulation.

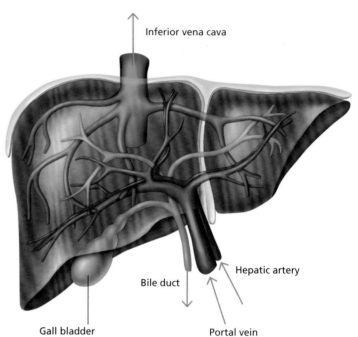

ABOVE **The liver processes nutrients and toxins delivered from the digestive tract through the portal vein, before they enter the rest of the body through the inferior vena cava. The hepatic artery mainly supplies the liver with oxygen and not nutrients.**

Phases of detoxification

The liver's ability to deal with poisons absorbed through the intestines is usually described as a two-phase process that is mediated by enzymes. In the first phase, enzymes (often belonging to a family called cytochrome P450) modify the structure of the compound, usually by introducing oxygen atoms to the molecular skeleton. This is then followed by the second phase, where other enzymes called transferases attach compounds such as glucuronic acid (present in cartilage and other connective tissues, for example) to the introduced oxygen atoms, thus increasing water solubility and enabling excretion of the neutralized toxin by the kidneys.

This detoxifying system is usually effective, but some poisons are hepatotoxic, that is they are directly toxic to the liver after absorption. The most famous examples of hepatotoxins are probably the amatoxins present in the destroying angel (*Amanita virosa*) and the death cap (*Amanita phalloides*) mushrooms. However, as these are fungi they will not be further treated in this book. Other instances of hepatotoxicity occur as a result of the chemical reactions taking place during the first phase. Some intermediary compounds produced can be toxic to the mitochondria of the liver cells, or to the enzymes involved in the many processes that the liver performs, including detoxification. They can also damage cells within the liver that have particular functions – for example, damage to the cells of the bile duct leads to a painful build-up of bile in the liver. The consequences of damage to the liver will be explored for various plant hepatotoxins, including coumarins (see pages 170–171) and pyrrolizidine alkaloids (see pages 164–167).

KIDNEYS

The kidneys are responsible for ridding the body of water-soluble waste products and toxins. Humans have two kidneys, one on each side of the body just below the ribcage, and each the size of a fist. They filter the blood, maintain the balance of electrolytes such as sodium and potassium, and are important for regulation of the blood pressure. The kidneys also make the active form of vitamin D, which is essential for healthy bone strength, and a hormone that stimulates production of red blood cells by the bone marrow.

Waste products and excess water filtered through the kidneys collect as urine in the bladder, from where they are excreted. The rate at which waste products are removed depends on the efficiency of the filtration, which will be affected by how well the kidneys are working and the blood pressure, as well as fluid and salt intake. Some compounds are nephrotoxic, or toxic to the kidneys. One example, the aristolochic acids in species of *Aristolochia* (see pages 168–169), can cause kidney disease and are also mutagenic, increasing the risk of cancer of the bladder and urethra when consumed over several years.

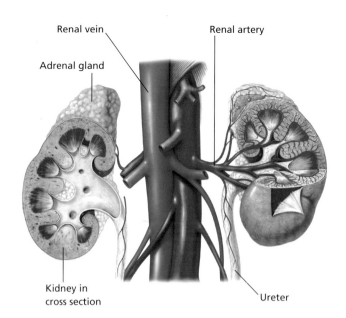

ABOVE **Anatomy of the kidney. Blood enters the kidneys through the renal artery and filtered blood leaves through the renal vein. Excess water and waste leaves via the ureter as urine. The adrenal glands at the top of the kidneys produce hormones such as adrenaline.**

Detox by every cell

It is not only the liver that produces enzymes that can detoxify plant poisons. For example, animals produce minute amounts of cyanide ions, which are extremely toxic. To cope with this hazard, animal cells contain the enzyme rhodanase, which converts cyanide to the far less toxic thiocyanate ion. Thiocyanate is then excreted in urine. Plants that produce cyanide-releasing compounds often do so in amounts that overwhelm the animal's ability to detoxify them should they be eaten, and are thus usually effectively protected against herbivores (see pages 182–185). However, some animals, such as the golden bamboo lemur (*Hapalemur griseus*) of Madagascar, feed on cyanide-containing plants and are immune. In most cases the mechanism for such immunity is unknown.

LEFT **The destroying angel (*Amanita virosa*) contains amatoxins, which are extremely toxic to the liver.**

ABOVE **The golden bamboo lemur (*Hapalemur griseus*) can eat bamboo that contains cyanogenic glycosides without being poisoned. It can withstand about 12 times the lethal amount for an animal its size.**

Heart and circulation

A beating heart is perhaps the primary sign of life, but what is its role in the body? The heart is a pump made from muscle, which causes blood to circulate around the body in blood vessels. In humans, as in other mammals, it is formed from four chambers: the two upper chambers, called the left and right atrium (plural atria), collect blood; and the two lower chambers, the left and right ventricle, pump the blood. This construction enables the blood to flow in two loops. The small (pulmonary) loop goes from the right ventricle to the lungs, and then back to the heart through the left atrium. The large (systemic) loop goes from the left ventricle out into the body, and then back to the heart through the right atrium.

In the pulmonary loop, the blood flows through the walls of the lungs, where it loses carbon dioxide and picks up oxygen. This 'oxygenated' blood then travels back to the heart, which pumps it through the arteries and around the rest of the body, including to the brain. Blood loses oxygen and picks up carbon dioxide as it moves around the body. The 'deoxygenated' blood returns, via the veins, to the right atrium at the start of the pulmonary loop, and the cycle then starts again. At rest, it takes around 60 seconds for a red blood cell to complete this journey through the circulatory system.

Pumping life

The heart consists of specialized muscle tissue. Contraction of the heart muscles is brought about by the shortening of the individual cells that make up the bulk of the muscle. Their contraction is coordinated by the cardiac pacemaker cells, which first trigger the atria to push the collected blood into the ventricles, and then trigger the ventricles to contract and force the blood out into the lungs and body, respectively. As the heart muscles relax, blood can flow into the atria in preparation for the next contraction. The pacemaker cells can act independently of the rest of the body's nervous system as they are capable of generating the triggering signal themselves. However, they also receive signals from other nerves, which enables the body to react in a coordinated manner to changes in circumstances. For example, a normal resting heart rate is approximately 70 contractions, or 'beats', per minute but can rise significantly during physical exertion.

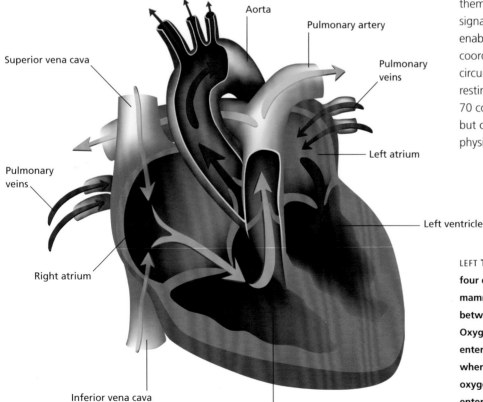

LEFT The human heart is divided into four chambers, like the hearts of other mammals and birds, with valves in between preventing backflow. Oxygen-poor blood from the body enters the right side of the heart, from where it is pumped to the lungs. After oxygenation, the blood from the lungs enters the left side of the heart and is pumped to the rest of the body.

Blood vessels

Arteries carry blood away from the heart under high pressure, and veins carry blood back to the heart under lower pressure. The fine vessels connecting arteries and veins in the tissues of the body are called capillaries. These are permeable and allow fluid from the blood to leak or filter out into the surrounding tissue at the arterial end and be reabsorbed at the venous end. It is through the capillaries that the exchange of oxygen and carbon dioxide, as well as nutrients and other waste products, takes place.

The blood vessels are also involved in the regulation of blood pressure and body temperature. Muscles in the walls of the vessels regulate their diameter, and by constricting will increase the resistance of the flow and thus raise the blood pressure. Some of the muscles in the capillary network can actually close a vessel off completely. This regulates how much blood flows through a tissue and is often easily observable, for example as pale white hands if you do not wear gloves on winter days, or as the flushed red face of someone who has just run up a flight of stairs. In the first case, the capillaries in the skin contract to prevent loss of heat, and in the latter, all capillaries are fully dilated to cool the body down.

Toxic interference

There are a number of ways in which toxins can interfere with the proper functioning of the circulatory system. Alteration to the nerve signals or direct effects on the heart muscles can change the force, frequency or regularity of the heartbeat. Toxins can cause blood vessels to contract, resulting in dangerously high blood pressure or making blood flow impossible and effectively starving the body's cells of oxygen and nutrients.

ABOVE **The odollam tree (*Cerbera odollam*; see page 53), like many members of the dogbane family (Apocynaceae), contains cardiac glycosides of the cardenolide type, toxins that affect the heart.**

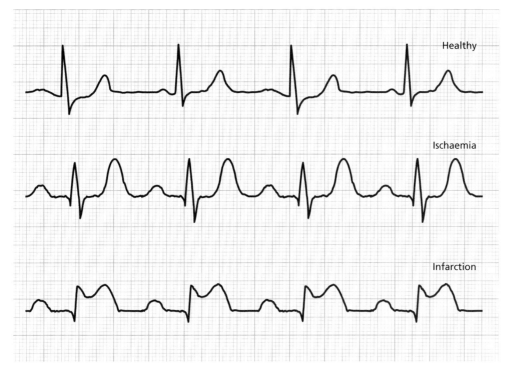

LEFT **An electrocardiogram (ECG) shows the changes in electrical activity during the contractions of the heart. With an ECG it is possible to monitor and measure the number of beats, and discover deviant activity due to toxins or disease.**

Brain and nervous system

Although we are aware of and can direct some of our actions, there are many other activities and processes that take place unnoticed. Here we look at the nervous system, which is responsible for consciousness, perception and regulation, and introduce how different aspects might be affected by plant poisons.

LEFT Seedpods of the South American vilca tree (*Anadenanthera colubrina*), whose seeds contain tryptamins such as bufotenin and are smoked in pipes and used to make a mind-altering snuff.

Consciousness

As thinking beings, we believe we are in control of our actions, but this is not necessarily the case. Our conscious mind is constantly going over past events or planning for the future, or is occasionally in the present moment. This thinking can be accompanied by movement, but we are usually not aware of initiating or controlling these actions. We are also receiving information from our senses, both external (such as sight, hearing, smell, taste and temperature) and internal (including hunger or the need to urinate). Some of this information is received, processed and acted upon without us being conscious at any stage, while at other times we have a vague awareness or make an active decision.

We are able to act in all these ways because of the complexity of our brains and the nature of our body's nervous system. The brain is divided into regions that have different functions. For example, we experience pleasure when a region of the brain is stimulated by particular compounds, such as dopamine. A substance that increases or reduces the production or retention of dopamine in this region will therefore affect our mood. Other regions control our breathing rate and depth, movement or speech, or enable concentration or planning, to name but a few of their roles.

Nervous system

The body's nervous system is formed of specialized cells called neurons. The neurons in the brain and spinal cord make up the central nervous system (CNS). This is connected to the rest of the body by the peripheral nervous system (PNS), which is formed of sensory neurons that relay messages to the CNS and motor neurons that relay messages from the CNS to the rest of the body. All these neurons communicate with each other through compounds called neurotransmitters. We have already mentioned one of these: dopamine. Other important transmitters that are affected by different plant poisons, as we will see later in the book (in particular, see Chapters 4 and 5), include noradrenaline (norepinephrine), serotonin, acetylcholine and *gamma*-aminobutyric acid (GABA).

The motor neurons of the PNS can be divided into the somatic (or voluntary) nervous system and the autonomic nervous system. There are two branches to the autonomic nervous system, which together control all the functions of the body that are outside conscious control (that is, they are automatic):

✦ the sympathetic nervous system, which controls the 'fight or flight' response to perceived danger, such as by speeding up the heart rate; and
✦ the parasympathetic nervous system, which controls the functions that enable the body to 'rest and digest', including slowing the heart rate.

The somatic nervous system, the other branch of the PNS, is mainly under our control and relays messages from the brain and spinal cord, principally to the muscles that enable us to move.

RIGHT **Parts of the nervous system: the brain; spinal cord (including a cross section); individual nerve cell with a long axon and shorter dendrites; and a synapse, where the signal is transmitted from one cell to another.**

Neuronal cells

Neurons come in a multitude of forms but can usually be recognized by their numerous connections with other cells, particularly other neurons. As with all cells, they have a body containing the nucleus, but in addition they have extensions called axons and dendrites, which are responsible for communication with other cells. Axons transmit signals to other cells and dendrites receive signals from other neurons. Motor neurons, which are responsible for transmitting the signals from the CNS to our muscles, can have very long axons, while some neurons in the CNS show little difference between axons and dendrites. The end of the axon branches and forms connections, such as with the dendrites of other neurons. Rather than making physical contact, however, the axon releases a neurotransmitter that can cross the small gap between the two (called the synapse) and binds to the receptor on the adjoining dendrite to activate the next neuron.

Neurotoxicity

It is the role of transporter proteins to remove, or enzymes to break down, the neurotransmitter so that the nerve cell is not constantly activated, and to allow another signal to be sent to the same receptor. If a toxin interferes with the enzyme that breaks down the neurotransmitter, it will not be removed but rather will build up, leading to prolonged interaction and overstimulation of the nerve cell. Other ways in which poisons can affect the nervous system are by masquerading as a neurotransmitter, or simply blocking the receptor and inhibiting the nerve signal.

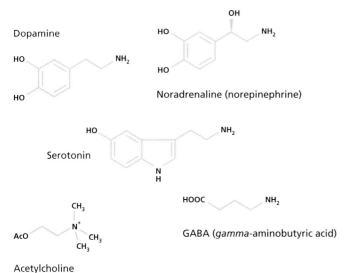

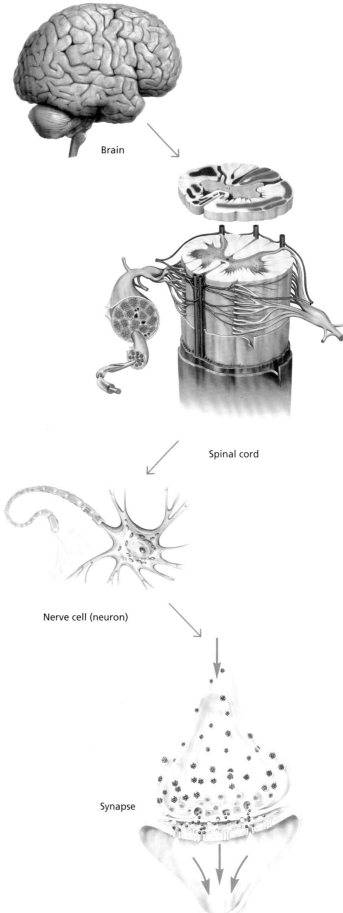

ABOVE **The main transmitter substances, or neurotransmitters, in the human nervous system.**

Muscles

When we move, we are aware of our muscles: the skeletal muscles that cause that movement, and the cardiac muscles of our beating heart. But those are not the only muscles we have. There is a third type, the smooth muscles, which are just as important but largely out of our control.

Types of muscle

Skeletal muscles are so called because they are attached to our skeleton and by their contraction and relaxation they cause it to move. These contractions are under our control and initiated by signals from our CNS (see pages 38–39). This control is finely tuned to the required movement, whether it is strong, fast contractions of leg muscles when we run, or the softer, intricate movement of fingers when we type. Some muscle movement, called reflex movement, is out of our control. An example is shivering when we are cold.

Cardiac muscle is found only in the heart. Its contraction, the heartbeat, is responsible for the movement of blood around the circulatory system (see pages 36–37). Although the heart muscles contract independently from the rest of the nervous system, they also respond to signals from the autonomic nervous system. These increase the speed and strength of the contractions for 'fight or flight', or slow them down to a base rhythm for 'rest and digest'.

Smooth muscles are not consciously controlled and respond to signals from the autonomic nervous system or from hormones. They include the muscles of the digestive tract (see pages 32–33) and are also found around the alveoli of the lungs, around blood vessels, in the eye and elsewhere. As previously indicated (see pages 36–37), smooth muscles contribute to regulation of body temperature by controlling how much blood circulates near the surface of the skin, but they also do this by causing goosebumps, which are actually the contraction of tiny muscles that cause hairs to stand erect and trap air. Another example of the role of smooth muscles, this time in the iris, is in the control of how much light enters the eye by regulating the size of the pupil.

How muscles work

There are differences in the structure of the cells that make up the three types of muscle. For example, in skeletal muscle individual cells fuse together and form long units that seem to have multiple cell nuclei; cardiac muscle cells are branched and have a single nucleus; and smooth muscle cells are unbranched and have a single nucleus. To function, they all need a good supply of oxygen for mitochondrial energy production in the form of ATP (adenosine triphosphate). Muscle contraction requires vast amounts of ATP and also depends on calcium ions, which are stored in an organelle called the sarcoplasmic reticulum. The release of calcium ions is regulated by the flow of sodium and potassium ions across the cell membrane, which in turn is usually initiated by nerve signals.

Skeletal muscles can continue to work for short periods without oxygen by using anaerobic processes, but this leads to a build-up of lactic acid and after a while they have to rest in order to remove that by-product. It is lactic acid that causes muscles to hurt after exertion. This alternative process cannot be used by cardiac muscle, as the necessary rest to remove lactic acid would mean that the heart would stop beating. Thankfully, cardiac muscles do not get tired as they have a very large number of mitochondria and a good blood supply that ensures constant provision of oxygen for their energy production.

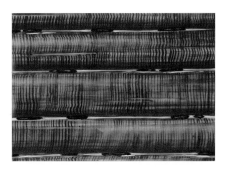

LEFT **Skeletal muscle:** long, striated fibres formed of a number of cells fused together and therefore with multiple cell nuclei.

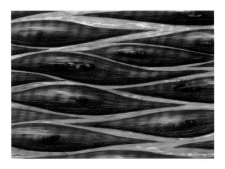

LEFT **Smooth muscle:** non-striated fibres of individual unbranched, spindle-shaped cells, each with a single nucleus.

LEFT **Cardiac muscle:** striated fibres of individual branched cells, each with a single nucleus.

ABOVE The tropane alkaloids in angel's trumpets (*Brugmansia* spp.) can cause dilation of a single pupil (unilateral mydriasis) if the plant juice gets into only one eye, such as when gardening or if the eye is rubbed after touching the plant.

BELOW Eyes with pinpoint pupils (top), or miosis, due to contracted muscles in the sphincter of the iris; and large, dilated pupils (bottom), or mydriasis, from inhibition of the iris sphincter and contraction of the radial iris muscles. Pinpoint pupils are caused by opiates such as morphine, while tropane alkaloids such as atropine can cause dilated pupils.

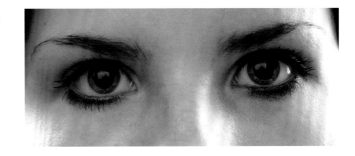

Effects of toxins on muscles

There are several ways to disrupt muscle function, and plants seem to have evolved a number of them. Plants may contain compounds that will disrupt the nerve signals to skeletal muscles, causing either seizures of painful and paralyzing contractions (see Chapter 4), or muscle relaxation and flaccid paralysis (see Chapter 5). Other plants contain substances that affect the smooth muscles, where poisoned humans may have the telltale signs of widely dilated pupils (see pages 80–83) or barely noticeable pinpoint pupils (see pages 200–201). Smooth muscles are common throughout the gastrointestinal tract, and some plant compounds can, for example, diminish the release of saliva, causing a dry mouth, or alter the time it takes food to pass through the intestines (see pages 144–145). Among the most potent plant toxins are those that will cause abnormal heart rhythm (see Chapter 3), either decreasing the number of beats per minute or causing irregular contractions, effectively hindering the circulation of the blood.

Endosymbiosis and sequestration

We have learnt that, according to the Darwinian theory of evolution, there is a constant competition for survival. Plants produce compounds that provide protection, while animals, fungi and bacteria develop ways of evading these toxins. But it is not always that straightforward; sometimes, a kind of collaboration evolves.

From infection to advantage

Plants are susceptible to infection by microbes just like animals (see pages 88–89), some of which can destroy entire crops and are feared agricultural pests. The Great Famine in Ireland in the middle of the nineteenth century was, to a large extent, due to a fungus-like organism causing potato blight, while wheat fields can be devastated by rust fungi and rice paddies spoilt by bacterial rot.

Just as we humans have bacteria in our guts and on our skin, some plants have developed intricate relationships with bacteria and fungi. In certain plant groups, notably the legumes, the roots form nodules that house bacteria capable of extracting nitrogen from the air and converting it to ammonia, which can then be used by the plant. The most extreme collaboration is probably the lichens, consisting of a symbiotic organism of intermingled fungal cells and photosynthetic algae. In this case, the fungus provides protection, and can even make the lichen poisonous, and is rewarded by the sugars from photosynthesis. Other fungi form mycorrhizae, mutually symbiotic connections to the roots of larger plants, making the uptake of water and minerals more efficient.

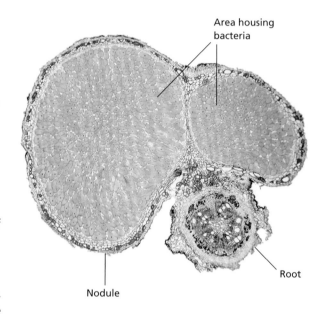

ABOVE **Microscopic cross section of soyabean (*Glycine max*) root nodule, showing the nitrogen-fixing bacteria (*Bradyrhizobium* sp.) stained yellow and the soyabean vascular tissue stained orange.**

Growing inside

There are many instances in which fungal species have been isolated from inside plant tissues. In most cases the role of these endophytes is unknown, but sometimes the fungi are responsible for the toxicity of the plants in which they grow. This can, for example, be due to the fungus producing toxins of its own (see pages 88–89) or metabolizing certain plant compounds (see pages 170–171).

In those instances where the poisonous compound in a plant is of interest as a human medicine or an insecticide, it is often of great value to find an endophytic fungus that produces the compound. This is because it is usually much easier to grow large amounts of fungi rather than plants and thus produce

LEFT **Soyabean (*Glycine max*) plant, in the pea or legume family (Fabaceae), with roots exposed to reveal the numerous root nodules that house nitrogen-fixing bacteria.**

the compound in sufficient quantity. This has been done with varying success for a range of plant-derived drugs such as paclitaxel (originally from the Pacific yew, *Taxus brevifolia*), camptothecin (originally from xi shu, or the happy tree, *Camptotheca acuminata*) and vincristine (originally from Madagascar periwinkle, *Catharanthus roseus*), all of which are used to treat cancer. In most cases, the fungus seems to have retrieved the genes necessary to make the enzymes producing the compound from the host plant.

Sequestering plant poisons

A silent chemical war rages between plants and herbivores. The evolution of a powerful deterrent or poison is clearly advantageous to plants, but the process is usually slow and gives herbivores time to develop resistance to the toxin. Such defence against poisoning can be due to any, or a combination, of several mechanisms. Earlier, we learnt about the rabbit's ability to break down atropine by producing a specific enzyme

(see page 23). Other defence mechanisms may involve an inability to absorb the toxin, or having transporter proteins that push the poison back into the intestines for elimination. There can also be structural differences at the poison's site of action. For example, insects are usually very sensitive to nicotine from tobacco. The structural differences between the insect receptors and the human receptors that bind nicotine mean that while the chemical is a deadly insecticide, it results in only a minor stimulating effect in humans at corresponding doses.

Among insects, not only has resistance to plant poisons evolved, but some insects can even 'sequester' the toxin and use it as protection against predators. This is known from several groups of insects: leaf beetles feeding on pyrrolizidine alkaloid-containing plants in the Asteraceae family become toxic to predators; moths resistant to cyanogenic glycosides can cause poisoning of birds; treehoppers can sequester scopolamine by sucking the sap of angel's trumpets (*Brugmansia* spp.); and sawflies have been known to contain alkaloids from white hellebore (*Veratrum album*). The most famous example of sequestration is probably the monarch butterfly (*Danaus plexippus*), which feeds on the cardenolide-containing milkweeds (*Asclepias* spp.) and thus becomes unpalatable and poisonous to birds (see page 53).

ABOVE **The treehopper *Alchisme grossa* guarding eggs in the Amazon rainforest of Ecuador. Treehoppers are known to sequester alkaloids from angel's trumpets (*Brugmansia* spp.), thought to be a defence against predators.**

LEFT **The alkaloids of the white hellebore (*Veratrum album*), from the mountains of Europe and northern Asia, are sequestered by the larvae of the hymenopteran sawfly *Rhadinoceraea nodicornis*.**

CHAPTER 3

MATTERS OF THE HEART

Most people regard a rhythmic heartbeat as the ultimate sign of life. Plants have, through evolution, developed several compounds that can affect how the heart performs its function of pumping blood around the body. This chapter looks at the plants and toxins that threaten the rhythm of the body's most vital organ.

CARDIAC POISONS: MECHANISMS OF ACTION

There is more than one way to disrupt or stop the beating of a heart, and plants produce compounds that exploit a variety of mechanisms with this end result. All the toxins in this chapter, which are either diterpene alkaloids or glycosides derived from steroidal lactones, directly affect the nerves and muscles of the heart.

ACONITE ALKALOIDS

Found primarily in the buttercup family (Ranunculaceae), these compounds (see pages 48–49) are formed by the introduction of nitrogen into a diterpene core structure. They act on the voltage-gated sodium channels in the membranes of cells that propagate nerve signals, the 'excitable cells'. In response to a nerve signal, these channels open and sodium ions (Na^+) are flushed into the cell, while negatively charged ions are forced out of the cell. The changes in ion concentrations on either side of the membrane will eventually close the sodium channels in the activated cell and open those of the next cell, thus passing on the nerve signal. At the target, the signal is transformed, for example, into a muscle contraction or the release of hormones.

Aconite alkaloids prevent these sodium channels from closing, thereby making it impossible for the excitable cell to reset and send on the next nerve signal. In skeletal muscle, this will lead initially to tingling sensations and, when a majority of sodium channels are affected, to paralysis. In cardiac muscle, it will disrupt the coordination of contractions, causing arrhythmias and eventually leading to failure of the heart to pump blood.

TAXUS ALKALOIDS

The alkaloid compounds present in the yew genus (*Taxus* spp.) are based on diterpenes, with the nitrogen atom incorporated as a sidechain attached to one of a number of core structures. The group of alkaloids originally isolated as a crude extract from the European yew (*T. baccata*) are collectively known as taxine, with taxines A and B being the most common and abundant alkaloids in several species. There are more than 400 compounds described, but due to its use as an anticancer drug the most famous of the *Taxus* alkaloids is probably paclitaxel. This compound is restricted to the Pacific or western yew (*T. brevifolia*), and is

BELOW **Diagram showing how the sodium–potassium pump moves intracellular Na^+ and extracellular K^+ ions over the cell membrane by splitting ATP into ADP and phosphate. This pump is inhibited by cardioactive glycosides.**

Sodium-potassium exchange pump
Extracellular fluid
Potassium ion (K^+)
Sodium ions (Na^+)
ATP
ADP
Phosphate
Cytoplasm
Intracellular fluid

RIGHT **The purple foxglove (*Digitalis purpurea*) from western Europe is a source of cardiac glycosides that have been used medicinally but are poisonous in larger amounts.**

present only in minute amounts in the inner layers of its bark. Thankfully, it is active against cancer cells at concentrations below those that adversely affect the heart.

Taxines bind to the same sodium channel as the aconite alkaloids, but in contrast to the latter they stabilize the closed form and are more selective for the sodium channels in cardiac muscle. As no sodium ions can pass through a closed channel, nerve signals are effectively terminated. When this occurs, the muscle contractions necessary for pumping the blood around the body can no longer be coordinated, leading to arrhythmias.

CARDIAC GLYCOSIDES

So named due to their action on the heart, cardiac (or cardioactive) glycosides occur quite widely in the plant kingdom as well as in some animals. They are mostly C_{23}-steroidal compounds derived from triterpenes. The two types of cardiac glycosides that occur in plants, the cardenolides and the less common bufadienolides, differ structurally (see pages 54–61) but have the same effect on the cells of the heart.

Cardiac glycosides inhibit the enzyme sodium–potassium adenosine triphosphatase (Na^+/K^+-ATPase), which uses the energy released by hydrolysis of ATP (adenosine triphosphate) to pump sodium ions out of cells and potassium ions (K^+) into them against their concentration gradients. High concentrations of K^+ within the cell and Na^+ outside the cell are necessary for normal signal transmission over the cell membrane.

ABOVE **Larkspurs (*Delphinium* spp.), such as this cultivated form, are popular garden plants that contain aconite alkaloids. Like aconites (*Aconitum* spp.), they are in the buttercup family (Ranunculaceae).**

Cardenolides and bufadienolides increase activity in the part of the central nervous system (CNS) responsible for slowing down the heart rhythm, resulting in fewer beats per minute. In binding to Na^+/K^+-ATPase in the membranes of cardiac muscle cells, they also produce a higher concentration of Na^+ inside these cells, which in turn results in the build-up of calcium ions (Ca^{2+}). The effect of this is to strengthen the contraction of cardiac muscles. These actions can be beneficial in low doses (see pages 198–199), but larger amounts can lead to heart failure.

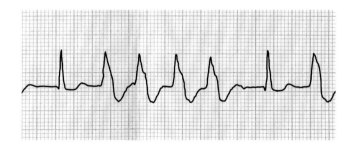

ABOVE **Electrocardiogram (ECG) showing 'extra beats' (ventricular extrasystoles) that may progress to fatal arrhythmias, which is one of the effects of aconite and *Taxus* alkaloids.**

Aconitum alkaloids

In addition to the general term aconite, species of *Aconitum* have a number of common names, including monkshood and wolfsbane. Aconite was described as a poisonous plant as early as the third century BC, when the ancient Greek philosopher and naturalist Theophrastus (c. 371–c. 287 BC) included it in his *Historia Plantarum*. The active principles are a group of alkaloids that are extremely toxic if eaten. These peculiar compounds exert their effect by interfering with nerve signals that are essential to the normal function of the mammalian body. Symptoms include a tingling sensation or numbness to touch, muscle paralysis and changes to the heart rhythm.

Origin and chemical characters

PLANT:
Aconitum napellus L. and *A. ferox* Wall. ex Ser.

COMMON NAMES:
A. napellus – monkshood, helmet flower, aconite;
A. ferox – Indian aconite

FAMILY:
buttercup (Ranunculaceae)

TYPE OF TOXIN:
diterpene alkaloids (aconitine, pseudaconitine)

SYMPTOMS OF POISONING IN HUMANS:
CIRCULATORY: arrhythmia
NEUROLOGICAL: tingling, numbness, weakness, paralysis

Aconite alkaloids are mostly restricted to a small number of genera in the buttercup family, particularly the aconites (*Aconitum* spp.) and their close relatives the larkspurs (*Delphinium* spp.). The presence of the compounds seems to give these plants a strong evolutionary advantage, as the group constitutes about a third of all species within the family. Aconites and larkspurs produce these highly toxic compounds from a substance called geranylgeranyl diphosphate, which is an essential part of the chlorophyll needed for photosynthesis. They can be grouped into three different 'flavours', called veatchine, atisine and aconitine alkaloids. The compounds in the last of these groups are the most toxic, which is thought to be due to their ability to pass through fat-containing barriers such as cell membranes and also the skin. This explains why gardeners and florists who, with bare hands, handle the cut stems or crushed material of aconite and larkspur plants in large amounts or for extended periods of time may experience mild symptoms of tingling or numbness.

LEFT **Wolfsbane (*Aconitum vulparia*, syn. *A. lycoctonum* ssp. *vulparia*) is a European species with yellow flowers. Its common name comes from its former use as a poison to kill wolves.**

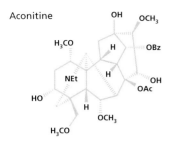

Aconitine

LEFT **The aconitine alkaloids are responsible for the potentially lethal heart arrhythmias caused by aconite poisonings.**

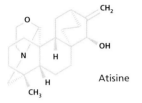

Atisine

LEFT **The atisine alkaloids cause numbness and tingling sensations, probably exerting their effects through the mammalian nervous system.**

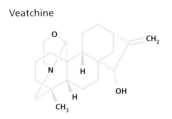

Veatchine

LEFT **Very little is known about the activity of the veatchine alkaloids, but they are important as starting compounds for other types of alkaloid.**

ABOVE **There are around 250 species of aconite (*Aconitum* spp.) found in the wild in the northern hemisphere, but they are also widely grown in temperate gardens and sold as cut flowers.**

Medicinal use

Species of aconite are used to treat joint pain in traditional Chinese herbal medicine in the form of bath additives, rubs and ointments. Their alkaloids can be absorbed through the skin, where they act as local anaesthetics. They may also be taken orally to treat asthma, gastroenteritis and various tumours, or as a supportive and revitalizing tonic. How is this possible, when these plants are very poisonous and small amounts can cause dangerous, or even lethal, effects on the heart and respiratory muscles?

Before use, the raw material must be subjected to *pao zhi* (a detoxifying measure), which might include heating and/or soaking with the intention of ensuring maximal therapeutic efficacy with minimal adverse effects. During such processes, the toxic alkaloids are transformed into less harmful compounds, explaining why aconite use has persisted in spite of its high risk of poisoning. However, even though *pao zhi* is usually performed, a number of patients who take traditional Chinese aconite medicines are hospitalized each year due to poisoning.

Another, unrelated species used in traditional Chinese herbal medicine, the Japanese meadowsweet (*Spiraea japonica*), a member of the rose family (Rosaceae), also contains aconite alkaloids, but these belong to the veatchine group, the least toxic structural variant.

A deadly curry

In a spectacular case of a jealousy-driven poisoning in the United Kingdom in 2009, a woman added powdered Indian aconite (*Aconitum ferox*) to a curry dish that was subsequently eaten by her former partner and his new fiancée. Within ten minutes, the man experienced a tingling sensation in the mouth that progressed to numbness and muscle weakness. Within a couple of hours, the pair showed heart arrhythmias and had difficulty breathing. The man died of ventricular fibrillation, as his heart became unable to keep the blood circulating, while his fiancée was supported by assisted ventilation for two days and made a full recovery within a week. Analysis of the curry and a herbal powder found in the scorned woman's possession led to her conviction for murder.

Yew and your heart

Taxus was the Latin name used by Romans for the yew tree, so the choice by Carl Linnaeus of the same name for the toxic genus seems pretty obvious. However, the etymology of the word is particularly interesting in a book about poisonous plants. The Romans are believed to have taken the name from the Greeks, who constructed the word *toxikon*, meaning a poison or drug used on arrows, from their word for a bow (*toxon*). As yew wood has a reputation of being the best for making bows, we have, in a way, come full circle.

DEADLY NEEDLES

PLANT:
Taxus baccata L.
COMMON NAMES:
yew, English yew, European yew
FAMILY:
yew (Taxaceae)
TYPE OF TOXIN:
Taxus alkaloids (taxine B)

SYMPTOMS OF POISONING IN HUMANS:
CIRCULATORY: abnormal heartbeat
NEUROLOGICAL: dilation of the pupils, dizziness, weakness, coma
DIGESTIVE: abdominal cramping, vomiting

ABOVE **Taxine B is a diterpene alkaloid due to the presence of a nitrogen atom in its sidechain. It is found in many species of yew (*Taxus*) and is structurally similar to the cancer drug paclitaxel.**

BELOW **Underside of a yew (*Taxus baccata*) branch, showing the pale lower surface of its needle-shaped leaves and both immature green and mature red berry-like arils, each surrounding a single seed.**

The yew is one of the few plants in this book that is not a flowering plant, but rather a conifer. As is typical of most other conifers, yews are trees or shrubs with modified leaves that we call 'needles'. Their seed 'cone', however, is very different from a pine cone: it is single-seeded, and by the time each seed is mature it is surrounded by a cup-shaped berry-like aril.

Yews are very slow-growing and long-lived. Probably the oldest living yew tree in Europe can be found in St Cynog's churchyard in Defynnog, Wales, and is thought to be around 5,000 years old. There is also evidence that the genus itself is very old in evolutionary terms. *Paleotaxus rediviva* fossils from the Triassic, dating back some 200 million years, are recognizable as yews, as are those of *Taxus jurassica* from the mid-Jurassic, 140 million years ago.

Today, the *Taxus* genus includes 12 species found around the world, including much of Europe, north Africa, China, the Philippines and Sumatra, Mexico, the United States and Canada.

All parts of the yew, with the exception of the arils, contain taxine alkaloids. The toxicity is not decreased on drying, so hedge trimmings are as toxic as the plant itself. Some deer seem able to eat the foliage, and sheep have been

ABOVE **With an estimated age of 5,000 years, the Defynnog yew (*Taxus baccata*) in St Cynog's churchyard in Wales is thought to be the oldest living tree in the United Kingdom and the oldest yew in Europe.**

known to browse on it unharmed. However, other animals – including horses, cattle, dogs and humans – are poisoned if they eat the leaves or branches. Poisoning of livestock is usually prevented by ensuring that they do not have access to these trees.

Food for badgers

The non-toxic arils produced by yews are sweet and of a gelatinous consistency when ripe. Their scarlet or scarlet-orange colour makes them easily seen by birds, which eat them whole, including the seed. The seed then passes through the bird and is deposited elsewhere. Children attracted by the arils usually have the good sense to spit out the seeds, which are larger than the seeds of a grape, and so do not come to any harm.

Scientists at the Royal Botanic Gardens, Kew, observed European badgers (*Meles meles*) eating yew arils off the ground and even standing on their back legs to eat them off the trees. They wondered why the animals were not poisoned. Dung pits found nearby that were full of partially digested arils and apparently intact seeds provided a possible answer. But being scientists, they wanted to test whether the seeds were in fact undamaged by the badgers' digestive system. They used liquid-chromatography mass-spectrometry (LC-MS) analysis (which separates and measures the mass or weight of compounds) to determine the presence and concentrations of alkaloids in arils and seeds taken from the trees, and also in seeds that had passed through the badgers. They found that there was no difference in the concentration of the main toxins in the seeds before or after they had been eaten. They also confirmed that the arils contained no alkaloids. The appearance of the arils in the dung suggests that they and the seeds pass through the badgers' digestive system very quickly, and the analysis confirmed that the seeds remained undamaged during their passage.

A Pocket Full of Rye

English crime writer Agatha Christie (1890–1976) qualified as an apothecary's assistant in 1917 and worked as a pharmaceutical dispenser in both world wars. She was thus familiar with the drugs and poisons of the early twentieth century, and subsequently wove them into the plots of some of her 66 detective novels. *A Pocket Full of Rye* (1953) tells of the events at Yewtree Lodge, where the wealthy Rex Fortescue, his latest wife and the maid (who had been trained by the amateur sleuth Miss Marple) are all murdered, the first death being due to taxine poisoning. In this twisting story of hidden identities and generational vengeance, the murderer needs to solve the problem of the bitter-tasting yew toxins. To mask this telltale sign, the poison is mixed in English marmalade made of Seville oranges, whose peel already imparts a slightly bitter taste to the spread.

APOCYNACEAE

The dogbane family is one of the larger families of flowering plants, and is today considered to contain more than 5,000 species in 366 recognized genera, including those that have, at times, been placed in their own family, Asclepiadaceae. The almost globally distributed Apocynaceae (only northern regions lack native species) has adapted to almost all environments and contains a large diversity of plant forms.

Species grow as herbs, climbers and lianas, succulents or trees. The flowers are often showy or conspicuous in form or smell, and many species have evolved special structures for pollen dispersal, such as pollinia, coherent masses of pollen grains that are transferred to the next plant by sticking to insect pollinators. These structures are especially elaborate in milkweeds (*Asclepias* spp.), waxflowers (*Hoya* spp.) and their relatives, constituting a feature that allows easy placement of these plants within the family (although deciding on the actual genus and species can be quite difficult).

HUMAN IMPORTANCE

The large number of species and wide geographical distribution of the dogbane family makes it easy to understand why so many plants are used by humans. The showy, waxy flowers of frangipani (*Plumeria* spp.) have found a place as a constituent in Polynesian lei garlands, the fibres from dogbane (*Apocynum cannabinum*) have been used to make cloth and string, some species are used in religious rituals (for example, see page 68), and some genera, such as *Landolphia*, were briefly important as sources of rubber in the late nineteenth and early twentieth centuries. Several plants in the family have been used as arrow poisons (for example, see pages 55–57) or in traditional medicinal systems, and the Madagascar periwinkle (*Catharanthus roseus*) is the source of an important cancer drug (see pages 208–209).

TWO TYPES OF TOXICITY

It seems that most plants in the dogbane family are toxic to some degree, but the reason for this differs between groups of species. Some groups produce cardioactive steroids as the toxic principle (see pages 54–59), while others produce monoterpene indole alkaloids (see page 68). Accordingly, the family presents several toxidromes, the combined picture of symptoms in poisonings, with some presenting as acute heart failure with arrhythmias and others giving signs of detrimental effects on the nervous system – for example, seizures, paralysis and hallucinations. As members of the dogbane family are widely distributed and many produce fatal intoxications, the use of these plants in suicides and poisonings is not uncommon in certain regions of the world.

LEFT **Elephant vine (*Strophanthus amboensis*) is found from Zaire to Namibia and contains cardioactive steroids. The petals are fused to form a cup at the base and there are five spreading, elongated lobes.**

THE SUICIDE TREES

The family includes the genus *Cerbera*, containing a small number of trees often growing in the ecologically important mangrove zones of tropical coasts. One species from Madagascar, the Madagascar ordeal bean (*C. manghas*), has – as its common name suggests – been used there as an ordeal poison. And the odollam tree (*C. odollam*), also known as the pong-pong, is responsible for half of the plant poisonings and 10 per cent of all poisonings in the Indian state of Kerala, where it grows along the coast. It is the seeds of the latter that are used for murder as well as suicide, resulting in some 50 deaths a year in Kerala alone. Knowledge of the toxicity of this plant has spread widely as a result, with headlines in newspapers tagging it the 'suicide tree'.

ABOVE **An odollam tree (*Cerbera odollam*) growing in Thailand. Its immature green fruit each contain a single poisonous seed.**

LEFT **Fruit of the odollam tree with the outer skin removed are used decoratively throughout the world.**

THE MACHIAVELLIAN MONARCHS

The North American monarch butterfly (*Danaus plexippus*) is famous for its seasonal mass migrations between its summer habitats in the United States and Canada, and its winter habitats in Mexico and the southern United States. It is also famous for sequestering toxic cardenolides (see page 43), as the larvae feed on milkweeds, making them unpalatable to preying birds. The toxins persist after the metamorphosis of the caterpillar, providing protection for the butterfly, the adult form. This is a two-step evolutionary process, where the initial development of resistance to the toxins in the food plants has subsequently evolved into a system where storing these toxins protects the herbivore against predators.

This plant–butterfly system is also often used as an example of mimicry, whereby a species that is similar to a poisonous species is protected through predatory association. In the case of the monarch butterfly, its form of mimicry is called Browerian mimicry, and explains how non-toxic individuals in the species are protected by the presence of toxic individuals. So even if a larva does not feed on poisonous food plants, the butterfly that develops from it will be protected because the majority of individuals in the species will contain toxic cardenolides sequestered from the milkweeds eaten by their larvae.

BELOW **A distinctively striped caterpillar of the monarch butterfly (*Danaus plexippus*) feeding on milkweed (*Asclepias* sp.), which has umbels of fragrant flowers with reflexed petals.**

Cardiac glycosides – a forceful beating

Plants that contain cardiac glycosides are found around the world in both temperate and tropical climates, and humans have been exploiting and abusing them for centuries. Their natural roles are not entirely clear, but at least some provide the plants with protection against herbivores.

Heart on steroids

PLANT: widespread in flowering plants, occurring in several families, including the dogbane (Apocynaceae), plantain (Plantaginaceae), buttercup (Ranunculaceae) and asparagus (Asparagaceae) families

TYPE OF TOXIN: cardiac glycosides (cardenolides, bufadienolides)

SYMPTOMS OF POISONING IN HUMANS IF EATEN:
CIRCULATORY: arrhythmia, raised blood pressure, heart failure
NEUROLOGICAL: headache, weakness, confusion, coma
DIGESTIVE: nausea, vomiting, diarrhoea

SYMPTOMS OF POISONING IN HUMANS IF INTRODUCED DIRECTLY INTO BLOODSTREAM:
CIRCULATORY: heart failure within a few minutes

Cardiac glycosides are plant steroids and, like all steroids, are derived from the linear triterpene squalene. Different steroid compounds are used in both plants and animals to stabilize cell membranes, as hormones and in the production of vitamin D. In biological chemistry, a structural scaffold that, through small modifications, can play a multitude of roles is called a 'privileged structure'. One such naturally occurring privileged structure is a steroid core of three specifically interlocked six-membered rings connected to a five-membered ring. This probably explains why cardiac glycosides are present in otherwise rather distantly related plant families, in a case of what is known as 'convergent evolution'.

Structural features

Cardiac glycosides share a number of other features besides being steroids. They all have one of two ring structures connected to the steroid core at a particular point. If the ring contains five atoms, such as in digitoxigenin, the compound belongs to the group called cardenolides (see pages 56–59), and if it contains six atoms it is called a bufadienolide (see pages 60–61). Both groups occur in plants, but never together in the same species and rarely in the same family. Bufadienolides also occur in animals – their name comes from the toad genus, *Bufo*, whose species produce the venom from which the bufadienolides, such as bufalin, were originally isolated.

The ring of a cardiac glycoside is necessary for its activity, as is the presence of an alcohol function – in other words, a hydroxyl group (–OH) – with a particular orientation, at the opposite end of the steroid structure. In plants, this is the 'glycosylation site', where one or more sugar compounds are attached. These are not strictly necessary for activity, but they modify the potency of the compound and the duration of its

LEFT **Flowering plants of lily-of-the-valley (*Convallaria majalis*) from Europe and temperate Asia contain cardiac glycosides of the cardenolide type. Leaves in particular sometimes cause poisoning.**

ABOVE **Cardiac glycosides are based on a steroid core structure synthesized from the linear triterpene squalene. The presence of an unsaturated lactone ring and a hydroxyl group at opposite ends of the structure is necessary for activity. Cardenolides such as digitoxigenin have a five-membered lactone ring, while bufadienolides like bufalin have a six-membered ring.**

Vincent van Gogh's xanthopsia

Medicines based on foxglove (*Digitalis* spp.) were not only used for ailments of the heart in the late nineteenth century, but also in the treatment of epilepsy. It has been hypothesized that Dutch painter Vincent van Gogh (1853–1890), who periodically suffered from a recurring illness of unknown identity, was prescribed such medicines and that the side effect xanthopsia, or yellow-tinted vision, was a contributing factor to his vivid use of yellow colours, as exemplified in his paintings of sunflowers. Support for this idea can be seen in two van Gogh paintings, in which the artist portrayed his doctor holding foxglove plants.

BELOW ***Portrait of Dr Gachet*, painted by Vincent van Gogh in 1890, during the final months of the artist's life.**

effects. Arrow poisons based on cardiac glycosides often contain no more than one sugar molecule, giving a rapid distribution to the heart and a short duration of activity due to effective renal excretion. As the number of sugar molecules increases, duration of activity increases correspondingly.

The different cardiac glycosides are also absorbed to differing degrees by the intestines. For example, digoxin from foxgloves (*Digitalis* spp.; see pages 198–199) is almost completely absorbed, while absorption of the strophanthins from kombe (*Strophanthus* spp.; see pages 56–57) is only minimal. Absorption appears to depend on how lipophilic the compound is, which is its tendency to combine with, or dissolve in, lipids or fats. Lipophilicity is influenced by the sugar portion of the compound and the number of –OH groups on the non-sugar, or aglycone, portion.

FAR-REACHING EFFECTS

Cardenolides and bufadienolides exert their toxic effect on Na^+/K^+-ATPase, an ion pump common to many cell membranes (see pages 46–47). In plants, this pump is probably involved in regulating the amount of water in, and thus the volume of, the cell. This might be a defensive mechanism in plants, as dehydration of leaves, stems or shoots will make them unpalatable to herbivores, and closing off the tissues prevents the spread of fungi and other infectious agents.

In humans, the sodium–potassium ion pump is also prevalent in the intestinal walls and in the lens of the eye, which explains why acute poisoning often causes diarrhoea and why taking more than the ideal medicinal dose for more extended periods can affect colour perception and give everything a yellow tint (see box).

The next few pages describe some plants that produce cardenolides and bufadienolides.

Cardiac glycosides – cardenolides

The dogbane family (see pages 52–53) includes many members that produce cardenolides. Several of these are significant, as their toxic nature has been used by humans as arrow, ordeal and rat poisons, as well as for murder and suicide. They have also been exploited by a few insects that are themselves tolerant of the toxins and able to store them in their cells, conferring on them some protection from predation.

Arrow poisons

PLANT:
Strophanthus kombe Oliv.

COMMON NAMES:
kombe

FAMILY:
dogbane (Apocynaceae)

TYPE OF TOXIN:
cardenolides (ouabain (g-strophanthin)

SYMPTOMS OF POISONING IN HUMANS IF EATEN:
CIRCULATORY: arrhythmia, raised blood pressure, heart failure
NEUROLOGICAL: headache, weakness, confusion, coma
DIGESTIVE: nausea, vomiting, diarrhoea

SYMPTOMS OF POISONING IN HUMANS IF INTRODUCED DIRECTLY INTO BLOODSTREAM:
CIRCULATORY: heart failure within a few minutes

Among the African plants in the dogbane family that are used as arrow poisons, members of the genus *Acokanthera* from east Africa have particular historical significance. Somali arrow poison (*A. schimperi*) is thought to be the source of the root mentioned by Theophrastus in the third century BC as being used to deadly effect by Ethiopians on their arrows. However, the plants that have been described as the number one for weapon poisons belong to the genus *Strophanthus*.

Kombe (*Strophanthus kombe*) is a woody climber that can grow as a shrub but is more usually a liana that can reach a height of 10 m (30 ft) when supported by other plants. The fruit are pairs of hard follicles that diverge at an angle of 180 degrees or more. Each follicle can be up to 50 cm (20 in) long and 2 cm (3/4 in) in diameter, and contains 100 or more seeds.

The Europeans who first thought to explore the newly 'discovered' African continent from the middle of the fifteenth century were made all too painfully aware of the use by the native people of poison arrows. More than one expedition was prevented from even disembarking from their ships because of arrows fired from the shore. They also observed elephant hunting from their ships, reporting that the animals were

LEFT **Ouabain** is a rhamnose cardenolide glycoside present in kombe (*Strophanthus kombe*) and several other species in the dogbane family (Apocynaceae) that have been used as arrow poisons.

LEFT Humans have found that a number of different plants produce toxins, including cardiac glycosides, which when used on the tips of arrows and darts improve their effectiveness during hunting.

ABOVE **The seeds of climbing oleander (*Strophanthus gratus*) contain ouabain and are used by nomadic people of Gabon, Cameroon and the Central African Republic to prepare an arrow poison.**

brought down with poison spears – sure evidence of how powerful the toxin was. The secret behind these deadly missiles was not easily discovered. It was not until explorer David Livingstone's Zambezi expedition of 1858–1864 that the Scottish botanist John Kirk (1832–1922) was able to collect material of the plant known as 'kombe' from south of Lake Malawi.

We now know that more than one species of *Strophanthus* is used as an arrow poison in Africa, with kombe being used most often in the east, brown or hairy strophanthus (*S. hispidus*) in the drier areas of the southwest and spider tresses (*S. sarmentosus*) in the northwest. Brown strophanthus is said to be preferred in the drier west African countries because it is more easily cultivated and has the advantage of retaining its precious seeds within the fruit for longer. As *Strophanthus* fruits usually burst easily when ripe to allow the feathery 'flight equipment' of the seeds to transport them away on the wind, hunters must harvest the fruits of species other than brown strophanthus before they are ripe, when the seeds are less toxic.

Strophanthus plants are a good choice of arrow poison. Their particular cardenolides, the strophanthins, are only poorly absorbed from the intestines, so animals that are killed by this method are safe to eat. The toxins are, however, very active when introduced directly into the bloodstream and will kill the prey quickly. Once the animal is struck, the hunter does not have to follow it for long before it can be caught.

Many rituals and practices grew up around the possession and use of kombe. The elders of a tribe often controlled who could grow the plant and kept the secret of how to prepare the toxin for use on arrows and spears. During the colonization of Africa by European countries, it became illegal to grow or possess kombe plants or seeds. Since that time, use of kombe has declined dramatically; there are now far fewer animals to hunt, but *Strophanthus* species still deserve their place among the most poisonous plants in Africa.

BELOW **The seeds of kombe (*Strophanthus kombe*) and of other members of the genus are the most poisonous part of the plants and therefore usually used to make arrow poisons.**

Sri Lankan suicide tree

PLANT:
Cascabela thevetia (L.) Lippold (syn. *Thevetia peruviana* (Pers.) K.Schum.)
COMMON NAMES:
yellow oleander, suicide tree
FAMILY:
dogbane (Apocynaceae)
TYPE OF TOXIN:
cardenolides (thevetins)
SYMPTOMS OF POISONING IN HUMANS:
as for *Strophanthus kombe* (see page 56)

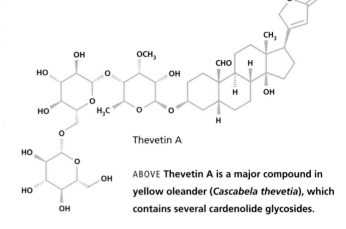

ABOVE **Thevetin A is a major compound in yellow oleander (*Cascabela thevetia*), which contains several cardenolide glycosides.**

Yellow oleander (*Cascabela thevetia*) is a large shrub or small tree with glossy green leaves and, as its common name suggests, yellow flowers. It is native to Mexico, Central America and northern South America, but is widely grown as an ornamental plant and has become naturalized in other countries with Mediterranean, subtropical or tropical climates. Among these is Sri Lanka, where the distinctively shaped seeds are frequently used for suicide attempts. This use seems to have been rare before 1980, but in that year an unusual case was published in a newspaper report and in subsequent years the 'popularity' of this means of attempting suicide increased at an alarming rate. There were thousands of cases a year by 2000.

Concerned by the number of patients it was receiving regarding yellow oleander poisonings, who were often dying during transfer from rural hospitals, the Institute of Cardiology at the National Hospital of Sri Lanka in Colombo decided to set up a study. This looked into whether methods used in Europe and North America for determining the severity of, and treating, poisonings from pharmaceutical digoxin could be applied in this case. They found that the assay used to measure digoxin levels in the blood does also detect other cardiac glycosides, but that it is an unreliable way to establish the severity of the poisoning. The treatment for pharmaceutical digoxin overdose, however, is also effective for treating poisoning by thevetins, and involves administering antibody fragments that bind digoxin. The main drawback is its cost, which makes it unaffordable as a course of treatment in Sri Lanka. Thankfully, most people who eat yellow oleander seeds make a full recovery if supportive care can be given.

Shared common names

Another plant in the dogbane family, *Nerium oleander,* is also known as oleander. This large shrub has grey-green leaves and pink, white or, occasionally, cream flowers that grow in small heads. It is native to the Mediterranean and east to Myanmar, but is also widely cultivated and is better able to tolerate colder climates than yellow oleander. Even though oleander is such a common plant, poisoning cases are comparatively rare, perhaps because of its relatively unattractive fruit and seeds. People have been poisoned by the plant after making tea from the leaves or eating the flowers, and the straight stems have also reportedly been used as poisonous kebab skewers.

Other heart-stopping families

Apart from the dogbane family, there are many other families with species that produce cardenolides. The foxgloves (*Digitalis* spp.), in the plantain family, are used as a source of important pharmaceuticals to treat weak heart function (see pages 198–199). In the buttercup family, the pheasant's eyes (*Adonis* spp.) are used in herbal medicine in Europe, North America and elsewhere. Cardenolides are also found in lily-of-the-valley (*Convallaria* spp.) in the asparagus family, and have caused poisoning when the leaves of this plant have been mistaken for wild garlic (*Allium ursinum*). Other plants in this family contain the second 'flavour' of cardiac glycosides, bufadienolides, and are covered overleaf. And then there is the deadly upas tree, in the mulberry family (Moraceae).

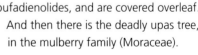

LEFT **Yellow oleander (*Cascabela thevetia*) has bright yellow flowers (left), and its angular fruits each contain a distinctively shaped seed (right).**

Deadly upas tree

PLANT:
Antiaris toxicaria (J.F.Gmel.) Lesch.
COMMON NAMES:
upas tree
FAMILY:
mulberry (Moraceae)
TYPE OF TOXIN:
cardenolides (antiarin)
SYMPTOMS OF POISONING IN HUMANS:
as for *Strophanthus kombe* (see page 56)

Fruit flies and the ouabain paradox

Insects have the same type of sodium–potassium ion pump that is present in humans, and the cardenolide ouabain was used early on in experiments investigating Na^+/K^+-ATPase from both humans and insects. The compound was shown to inhibit the function of the enzyme effectively in both organisms, but while ouabain is a potentially deadly poison for humans, many insects are resistant and remain practically unaffected. This has been called the ouabain paradox. The mystery was not solved until molecular biology and genetic studies on fruit flies (*Drosophila melanogaster*; see photo) showed that they have another transporter enzyme, which is co-located with Na^+/K^+-ATPase. It is this second pump that prevents the concentration of the cardiac glycoside from becoming high enough to inhibit the sodium–potassium ion pump.

ABOVE The upas tree (*Antiaris toxicaria*), found in tropical forests such as those in West Kalimantan, Indonesia, has a poisonous sap used on darts and arrows. The prey is paralyzed but the meat is safe to eat.

The widespread upas tree (*Antiaris toxicaria*), whose distribution ranges from tropical Africa to Southeast Asia and the Pacific islands, is probably more notorious than it deserves to be. Its sap became known to Europeans early in the fourteenth century as a dart and arrow poison, but later accounts talked of its ability to harm animals and plants for kilometres around. It was said that birds perching in its branches became dizzy and fell down dead, and that the soil beneath the tree and for a stone's throw away stayed barren and scorched. It is now known that the toxins in the upas tree are cardenolides, and although their use as an arrow poison is true, the other tales are the product of active imaginations.

BELOW Antiarin is the major cardenolide poison from the upas tree (*Antiaris toxicaria*). It is a rhamnose glycoside and a very potent arrow poison, but absorption after ingestion is low.

Cardiac glycosides – bufadienolides

The second 'flavour' of cardiac glycosides is the bufadienolides. However, while the bufadienolides in plants are glycosides (containing sugars in their structure), the toad venoms contain amino acid esters, which determine their solubility and absorption-modifying activity. Bufadienolides are present in fewer plant groups than cardenolides, but they are prevalent in the Scilloideae subfamily of the asparagus family (for example, in squills, *Drimia* spp., and their relatives), as well as in hellebores (*Helleborus* spp.), from the buttercup family. Historically, they have been important as sources of medicinal drugs, but due to their toxicity they are seldom used in modern medicine. In a review of 924 cases of cardioactive poisoning, the mortality of cardenolides (which were involved in the majority of cases) was found to be 6 per cent, while that of bufadienolides was close to 30 per cent.

Fighting rats

PLANT:
Drimia maritima (L.) Stearn (syn. *Charybdis maritima* (L.) Speta, *Ornithogalum maritimum* (L.) Lam., *Scilla maritima* L., *Squilla maritima* (L.) Steinh., *Urginea maritima* (L.) Baker)

COMMON NAMES:
sea squill, squill, red squill, sea onion

FAMILY:
asparagus (Asparagaceae)

TYPE OF TOXIN:
bufadienolides (proscillaridin, scilliroside, plus several others)

SYMPTOMS OF POISONING IN HUMANS:

CIRCULATORY: arrhythmia, raised blood pressure, heart failure

NEUROLOGICAL: coma

DIGESTIVE: nausea, vomiting, purgative diarrhoea

Sea squill (*Drimia maritima*) is one of the truly ancient drugs, and is described in the Ebers Papyrus, an Egyptian medicinal text from the sixteenth century BC. In the classical Greek texts by Theophrastus and the physician Hippocrates (c. 460–c. 370 BC), the plant was recommended for its diuretic properties and as a laxative, but also to treat convulsions and asthma. It was considered useful in warding off evil and protecting against spirits.

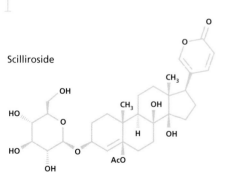

Scilliroside

ABOVE **Scilliroside from sea squill (*Drimia maritima*) is an example of a bufadienolide cardiac glycoside. It contains glucose and is also an acetic acid ester.**

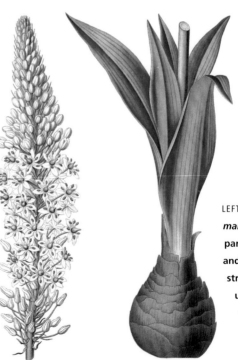

LEFT **Sea squill (*Drimia maritima*) is native to parts of southern Europe and north Africa. Its strap-shaped leaves have usually died back by the time the tall spikes of white flowers open in the autumn.**

In modern times, sea squill has been used as a rat poison. The cardiac glycosides it contains are chemically stable compounds that do not break down or lose their activity during drying. And although baits containing sea squill usually have a very bitter, unpleasant taste, deterring most animals, rats happily devour them and succumb to the toxicity. Rat poison was traditionally produced in large amounts by collecting sea squill bulbs, separating and cutting up the individual bulb leaves and drying the pieces, which could then be mixed with other bait. One reason for the efficacy of this rodenticide is that rats cannot vomit, which is otherwise one of the first symptoms of cardiac glycoside poisoning.

Between Scylla and Charybdis

In 1753, Carl Linnaeus chose the name *Scilla* for the genus containing sea squill, to which he gave the binomial *Scilla maritima*. A later taxonomist chose another species, the non-toxic alpine or two-leaf squill (*S. bifolia*) as the 'type' species that defines the genus. Unfortunately, however, modern studies have shown that sea squill is not closely related to this blue spring flower. As a consequence, sea squill is today placed in the genus *Drimia*. But it has taken a few attempts to get to this point, and among the synonyms we also find *Charybdis*, which like *Scilla* (*Scylla*), is the name of a sea monster. In the myth, Scylla and Charybdis protected a maritime passage, forcing seafarers – famously including Odysseus – to choose between two evils. With the confusing taxonomy of the genus *Scilla*, trying to figure out if a species is toxic can be similar to this choice between Scylla and Charybdis. However, if you can deduce that the plant you are looking at belongs to the modern concept of *Scilla*, you can be assured that its toxicity is very low.

Climbing onions

The strange-looking climbing onion (*Bowiea volubilis*; asparagus family) consists of a pale greenish bulb, up to 12–20 cm (5–8 in) in diameter, growing mainly above the soil surface. During the growing season, one or a few thin stems develop into long, winding, branched vines, forming a climbing, feather-like crown with small greenish flowers.

The climbing onion contains bufadienolides (bovisides) and has been used in the traditional medicine system of the Xhosa and Zulu peoples of South Africa to treat headaches, heart palpitations and infertility, and as a violent purgative, abortifacient and love-charm emetic. Even though many reports of deadly poisonings from climbing onion are known,

ABOVE **The large bulb of the southern African climbing onion (*Bowiea volubilis*) grows above the soil surface. Slender-stemmed, branching flowerheads grow quickly in spring or summer, reaching 6 m (20 ft) in length.**

it is still considered to be one of the most popular traditional drugs in South Africa. Due to this popularity, the species has declined in the wild and cultivation programmes have been established. The challenge now is to convince people that the cultivated plants are as effective as those collected from the wild.

CHAPTER 4

BREAKING THE BRAIN

As the nervous system, and especially the brain, is probably the most complex organ system in the human body, it is open to many different ways for poisons to cause deleterious and drastic effects. This chapter explores examples of poisons affecting the brain and other parts of the central nervous system (CNS).

CNS TOXINS: MECHANISMS OF ACTION

Some compounds work directly on our perception of reality, causing confusion or hallucinations, while others may affect normal nerve signalling, leading to seizures or paralysis. Certain compounds shortcut the nervous system control functions extinguishing life-saving reflexes that normally prevent asphyxiation or paralysis and muscle cramps.

INTERACTIONS AND INTERRUPTIONS

The nervous system consists of a complex, intertwined series of pathways, balancing the effects of several different neurotransmitters. As discussed in Chapter 2, we are aware of a small part of the activity in the brain, but most of its actions pass us by unnoticed. It is only when foreign substances act on receptors and enzymes, or even kill neurons, adversely affecting these unnoticed processes, that their importance becomes evident. These substances are collectively called neurotoxins, a term that covers everything from heavy metals like lead, to synthetic pesticides and drugs, and natural plant compounds.

One neuronal function we are unaware of is the brain's constant decision not to act on sensory input from the rest of the body. Through inhibitory pathways in the CNS, the brain keeps our muscles from constantly contracting as a reaction to external stimuli. Some plants contain compounds that affect these pathways, such as the alkaloids from nux vomica (*Strychnos nux-vomica*) and heartbreak grass (*Gelsemium elegans*) (see pages 66–69), which block glycine receptors in the spinal cord, leading to seizures. What can probably be considered the main inhibiting system in the brain is regulated by the *gamma*-aminobutyric acid (GABA) receptors, involved for example in sleep and muscle relaxation. There are several plant substances that bind to these GABA-receptors and cause muscle cramps, including sesquiterpenes from species of star anise, most notably Japanese star anise (*Illicium anisatum*), and tutus (*Coriaria* spp.) (see pages 70–73), and the polyacetylenic alcohols from the water hemlocks (*Cicuta* and *Oenanthe* spp.; see pages 76–77).

ABOVE **Engraving of nux vomica (*Strychnos nux-vomica*) showing a flowering stem (centre), and a flower, seed and fruit cut in half and whole (anticlockwise from top).**

RIGHT **The disc-shaped brown seeds of nux vomica (*Strychnos nux-vomica*) contain high levels of the poison strychnine.**

CONFUSION AND CONVULSIONS

Some neurotoxins also affect our consciousness and mental state. With the multitude of interconnections between cells and pathways in the CNS, other possible targets for neurotoxins include sodium channels. Although toxins disrupting these ion transporters are more famous for their effects on the heart (see Chapter 3), the grayanotoxins from the heather family (Ericaceae) have a greater affinity for the sodium channels present in the brain (see pages 78–79).

The classical witch's brew is considered to have been based on atropine and/or scopolamine, alkaloids present in several species from the potato family (Solanaceae; see pages 80–83). These alkaloids inhibit the effects caused by acetylcholine in the parasympathetic nervous system through muscarinic receptors. These receptors are widespread, and although many effects can be related to peripheral sites, inhibition of receptors in the CNS are responsible for causing agitation, hallucinations and convulsions.

TRANSMISSION OVERDRIVE

The mental state of a person is usually considered to be influenced by the transmitter substances noradrenaline (norepinephrine), dopamine and serotonin. From the pharmacological standpoint, increasing the activity of noradrenaline and serotonin affects depression, while drugs acting on the dopamine and serotonin systems are central in treating psychosis. Affecting these transmitter substances in the brain often leads to one of the most noticeable activities of neurotoxins: hallucinations. However, the effects are often very unpredictable, for example with the poison bulb (*Boophone disticha*) (see pages 84–85).

Among plant species containing monoterpene indole alkaloids, the iboga (*Tabernanthe iboga*) plant has been used in religious ceremonies to induce hallucinations in Africa (see page 68), while the peyote cactus (*Lophophora williamsii*) and yopo (*Anadenanthera peregrina*) have been used for the same purpose in North and South America, respectively (see pages 86–87). Iboga affects several pathways in the brain, including those of dopamine and serotonin, as well as receptors for opioids. The mescaline contained in the peyote cactus causes hallucinations by activation of serotonin receptors, but also shows activity in the dopamine pathway. Yopo contains the serotonin analogue *N,N*-dimethyltryptamine, more famous as the hallucinogenic constituent of the ayahuasca drink (see pages 86–87). The most famous hallucinogen is probably the synthetic lysergic acid diethylamide (LSD), which can be produced from alkaloids isolated from fungus-infected bindweeds and grasses. In the latter case, the compounds produce vasoconstriction instead of hallucinations (see pages 88–89).

ABOVE **Angel's trumpet (*Brugmansia suaveolens*) shrubs contain tropane alkaloids. The drooping trumpet-shaped flowers, up to 30 cm (12 in) long, are pollinated by moths, which they attract with their fragrance in the evening.**

LEFT **A shaman oversees ceremonies at which ayahuasca is drunk in tropical South America, including Ecuadorian Amazonia. The drink is made from a combination of plants, including ayahuasca vine (*Banisteriopsis caapi*).**

Monoterpene indole alkaloids – tonic or toxic?

One of the most diverse groups of compounds in plants is the monoterpene indole alkaloids (MIAs). These are synthesized from tryptamine and secologanin, producing the intermediate strictosidine. This compound is then converted to several thousand biologically active alkaloids, with strychnine probably being the most famous poison. Although MIAs are present in several plant groups, the related Loganiaceae, Gelsemiaceae, dogbane (Apocynaceae) and coffee or bedstraw (Rubiaceae) families are particularly rich in these compounds. Aside from several poisons (see also pages 94–97 and 136–137), some of the MIAs also include important medicinal drugs, used to treat diseases such as cancer or malaria (see pages 204–205 and 208–209).

BALANCING THE RESTORATIVE TONIC

PLANT:
Strychnos nux-vomica L.
COMMON NAMES:
nux vomica, strychnine tree, poison nut
FAMILY:
Loganiaceae
TYPE OF TOXIN:
monoterpene indole alkaloids (strychnine)

SYMPTOMS OF POISONING IN HUMANS:
NEUROLOGICAL: tingling, twitching, spasms, violent clonic convulsions, respiratory failure
DIGESTIVE: nausea, vomiting

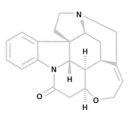

Strychnine

LEFT **The monoterpene indole alkaloid strychnine is a poison that historically has also been used as a reinvigorating tonic.**

Nux vomica is a shrub or small tree from India, the East Indies and Malaysia. Its small green flowers, which are said to have a foul smell, are followed by round, hard-shelled fruit up to 7.5 cm (3 in) in diameter. Although the fruit pulp is sometimes eaten, the disc-shaped seeds inside the fruit contain high levels of strychnine. This poison is also found in some other species of the genus *Strychnos*, usually in smaller quantities, such as in the St Ignatius bean (*S. ignatii*), from the Philippines, from which it was first isolated.

Strychnine is an odourless but bitter compound with a long history of use by humans. Its main use has been to poison animals that are seen as pests, and although the principal among these has been rats, it has also been used to kill wolves and other predators as recently as the 1970s. Historically, it has been used medicinally in small quantities in restorative and invigorating tonics to stimulate the appetite and senses, or in experimental treatments of people with peripheral paralysis, for example due to lead poisoning or polio.

This medicinal use is very risky, however, as poisoning from strychnine is truly horrific. The alkaloid is quickly absorbed through the lining of the nose or the digestive canal. It causes the skeletal muscles to contract at the slightest stimulation and to stay fully contracted for extended periods. The body is thrown into severe seizures of rigidity, often causing

BELOW **Dried branch of a nux vomica tree (*Strychnos nux-vomica*) from Thailand, with one of the hard, round fruits cut open to reveal two seeds held within the white flesh.**

opisthotonus, a state in which the body is poised in a backward arch. It also causes a particular facial expression called *risus sardonicus* due to contraction of the facial muscles (see page 77). These repeated convulsions eventually lead to death by respiratory arrest or exhaustion. The unfortunate victim is fully conscious, with heightened senses, throughout.

Monkey business

The fruit shells of *Strychnos* species, commonly known as monkey apples, are used to make decorative objects, such as containers for candles and salt and pepper pots, and as ingredients in potpourri. Luckily, the African species *S. spinosa*, which does not produce strychnine, is most commonly used for these objects; its main alkaloid is the far less toxic 11-methoxy-diaboline, and this is present only in low concentrations. Only one African species, *S. icaja*, is known to produce strychnine. In a fascinating twist, a number of species in the *Strychnos* genus, especially those found in South America, produce muscle-relaxant alkaloids used in curare (see pages 94–97) rather than cramp-causing strychnine.

The road to heartbreak

The small family Gelsemiaceae contains three genera (previously part of the Loganiaceae or borage (Boraginaceae) families) whose members are either small shrubs, climbing lianas or a substantial tree up to 40 m (130 ft) tall. The genus *Gelsemium*, with three species, all climbing plants with heads of yellow flowers, is the most well known. Two species – yellow or Carolina jessamine (*G. sempervirens*) and Rankin's or swamp jessamine (*G. rankinii*) – are found in North and Central America, and one species – heartbreak grass (*G. elegans*) – occurs from Indonesia and northwards into China. These plants contain more than 50 different MIAs, of which gelsemine and koumine usually have the highest concentrations. Gelsemicine and gelsenicine are the most poisonous of the *Gelsemium* MIAs, but in contrast to strychnine they initially cause muscle relaxation and sedation.

Heartbreak grass is used extensively in traditional Chinese medicine as an externally applied remedy for rheumatism, spasticity and headache. Doses of up to 50 g of plant material are considered safe when used topically, while as little as 2–3 g will be poisonous if ingested. The species' potentially lethal qualities have occasionally been used to harm rather than heal. In 2011, Chinese tycoon Long Liyuan was murdered over a business dispute; heartbreak grass had been added to his lunch of cat stew. Other cases of poisoning in China, however, have resulted from similarities in appearance between heartbreak grass and less toxic medicinal plants (see box).

RIGHT **Yellow or Carolina jessamine (*Gelsemium sempervirens*), also known as evening trumpet flower, is one of two species of *Gelsemium* from the southern United States, where it prefers dry upland habitats.**

Tragedy at the funeral dinner

The toxicity of heartbreak grass is largely unknown outside Southeast Asia, but in that region, particularly in China, it has caused numerous poisonings. In November 2011, a man brewed a bucket of medicinal liquor using the stems of a plant that he thought was sargent glory vine (*Sargentodoxa cuneata*) in the Lardizabalaceae family. When he drank some of the liquor later that day he had no symptoms, but a drink of the same liquor the next morning, 24 hours after it had been brewed, resulted in his death within an hour. The liquor was not known to be the cause of his death until later that day when his funeral dinner was held. Of the 34 people who attended the dinner, the ten who drank some of the medicinal liquor all became unwell, and four people died. Later analyses detected the toxin gelsemine in the liquor and stems of heartbreak grass were also identified. It seems that the concentrations of gelsemine and other toxins in the liquor increased gradually and had not reached a toxic level the first time it was drunk.

INITIATION BY IBOGAINE

PLANT:
Tabernanthe iboga Baill.
COMMON NAMES:
iboga
FAMILY:
dogbane (Apocynaceae)
TYPE OF TOXIN:
monoterpene indole alkaloids (ibogaine)

SYMPTOMS OF POISONING IN HUMANS:
CIRCULATORY: arrhythmias
NEUROLOGICAL: convulsions, hallucinations
DIGESTIVE: nausea, vomiting

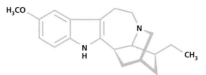

Ibogaine

ABOVE **The monoterpene indole alkaloid ibogaine has been used in religious ceremonies and is probably the best-known hallucinogen originating in Africa.**

The root, root bark, stem and leaves of iboga, a shrub in the dogbane family, contain iboga alkaloids, including ibogaine. Iboga grows in rainforests from west-central tropical Africa to Angola and is a sacred plant to the people of Gabon where it is used in the initiation rites of the Bwiti religion. It is one of only a few plants in the dogbane family that are known to have mind-altering effects, and is also unusual in that both adult men and women are permitted to take it; the use of most other traditional hallucinogens is restricted to men.

In the West, there are anecdotal reports that ibogaine can be used as a treatment for drug dependence, particularly of heroin and other opiates, with even a single dose leading to a sustained elimination of drug cravings. There are many reasons why this unorthodox treatment should not be attempted, including the fact that no safe dosage has been established, and side effects include gastrointestinal distress and interference with heart rhythm. Several deaths associated with ibogaine have also been reported.

LEFT **This iboga (*Tabernanthe iboga*) shrub in Limbe Botanical Garden, Cameroon, has heads of small, pale yellow flowers and orange fruits that taper to a point. It is the roots, and particularly their bark, that is used as a stimulant and during initiation rituals.**

Development of antidepressants

Indian snakeroot (*Rauvolfia serpentina*) is another member of the dogbane family, and is found from the Indian subcontinent to south-central China and west Malesia. It is traditionally used in Ayurvedic medicine as a hypnotic and sedative for the treatment of 'moon disease', or lunacy, and the Indian leader Mahatma Gandhi (1869–1948) is said to have chewed its root as an aid to meditation. The mechanism of action of the plant's main indole alkaloid, reserpine, includes a depletion of transmitter substances in the CNS, and frequent use will cause long-lasting depression. This unwanted effect turned out to be a significant discovery in the development of antidepressant drugs such as fluoxetine (Prozac), as their effectiveness could be assessed by their ability to prevent the depression caused by reserpine.

LEFT The roots of Indian snakeroot (*Rauvolfia serpentina*) have been used medicinally for centuries.

RIGHT Due to concerns over the sustainability of harvesting roots from wild populations for the pharmaceutical industry, international trade is now controlled.

HERBAL VIAGRA, WITH RISKS

Another famous MIA-containing plant is yohimbe (*Corynanthe johimbe*, syn. *Pausinystalia johimbe*), in the coffee family. The bark of this large tree, native to the region extending from Nigeria to west-central tropical Africa, contains yohimbine. Its action is a stimulant effect on the nervous system, and at low doses it also causes dilation of the peripheral blood vessels, which in some males causes priapism (a painful, enduring erection). Yohimbe bark has traditionally been used in central Africa as an aphrodisiac and is marketed globally as a 'food supplement', although this is prohibited in several countries, including the United Kingdom and Sweden. Other effects can include increased perspiration, hot and cold flushes, anxiety and seizures, and at high doses it can cause the blood pressure to spike. Several cases of myocardial infarction (heart attack) have also been reported.

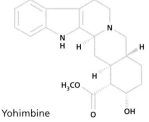

Yohimbine

LEFT Yohimbine is considered to be a stimulating monoterpene indole alkaloid and works mainly by blocking inhibitory peripheral nerve signals in the body.

RIGHT Although the sale of yohimbe (*Corynanthe johimbe*) bark as a 'food supplement' is prohibited in many countries, its availability online means it still poses a risk to consumers.

Seizure inducers – sesquiterpene lactones

As discussed on page 26, terpenes come in many sizes and forms. A very large group of terpenes that are hazardous to humans is the sesquiterpene lactones. The lactone ring is the reactive part of these compounds, and many of them irritate the skin (see pages 126–127), or are allergenic or even possibly carcinogenic. In this section we discuss only the small group of sesquiterpene lactones that cause potentially lethal seizures, such as those present in some star anise (*Illicium* spp.).

Star signs

PLANT:
Illicium anisatum L.
COMMON NAMES:
Japanese star anise, shikimi
FAMILY:
Schisandraceae
TYPE OF TOXIN:
sesquiterpene lactones (anisatin, neoanisatin)

SYMPTOMS OF POISONING IN HUMANS:
CIRCULATORY: arrhythmias
NEUROLOGICAL: malaise, tremor, tonic–clonic seizures, respiratory arrest
DIGESTIVE: diarrhoea, vomiting

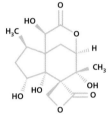
Anisatin

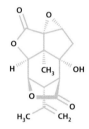
Picrotoxinin

ABOVE Some sesquiterpene lactones cause seizures by inhibiting GABA-receptors in the brain. Well-known examples include anisatin from Japanese star anise (*Illicium anisatum*) and picrotoxinin from fish berry (*Anamirta cocculus*).

BELOW The Japanese star anise (*Illicium anisatum*) is an evergreen shrub that bears star-shaped flowers in the spring. Its fruit are very similar to those of the Chinese star anise (*I. verum*), shown right.

The dried star-shaped fruit of the star anise or Chinese star anise (*Illicium verum*) tree is used in traditional Chinese medicine, herbal teas and as a flavouring in some cuisines, particularly those of China and Vietnam, where the species is native. The fruit's liquorice-like taste is due to the presence of high concentrations of anethole, a volatile simple phenylpropanoid.

Serious cases of poisoning, including epileptic reactions, have occurred when star anise has been unintentionally substituted with Japanese star anise in herbal teas. In the

early 2000s, this led to product recalls, import restrictions and public health warnings in several countries across Europe, Southeast Asia and North America. Morphological examination, chemical analyses and molecular techniques, which detect even small levels of contamination, should now ensure that the correct ingredient is used. Unlike star anise, Japanese star anise contains only small amounts of anethole. More importantly, however, it includes significant levels of toxic compounds belonging to a group of structurally unique sesquiterpene lactones found in the *Illicium* genus, such as anisatin and neoanisatin.

Japanese star anise is native to Japan, South Korea and Taiwan, and is cultivated in China and Vietnam. Known in Japan as shikimi, it is associated with Buddhist shrines and graveyards. The evergreen tree is planted in such places, and leafy sprigs are sold to visitors to place on graves, possibly as protection from wild animals, and to give as offerings to Buddha. Such evergreens are considered an appropriate offering as, unlike flowers, which soon fade and die, they can be kept alive simply by refreshing the water. The wood of the Japanese star anise is used to make incense, but perhaps the most intimate use of this fragrant plant is when branches are placed around the body of a recently deceased person to purify the air during their funeral service.

When the antidote is a poison

In 1903, Veronal, a new, effective and synthetic sleeping pill, was launched. It contained the first commercial barbiturate (barbituric acid derivative), barbital (sometimes known as barbitone), and would be followed by numerous others. Such drugs became very popular, but their use was plagued by lethal overdoses and addiction. They act through several systems in the CNS, causing profound depression of consciousness, blood pressure and muscle activity – including those muscles necessary for breathing. Before mechanical ventilation became common practice in the mid-1950s, several stimulating antidotes were experimented with. Cocaine, adrenaline (epinephrine) and strychnine had little effect on patients suffering from barbiturate poisoning. When picrotoxin was tried, it counteracted the muscle paralysis and kept the patient alive until they could metabolize the barbiturate. However, frequent overdosing of this antidote led to death anyway from seizures.

ABOVE **Despite their use as a fish poison, the dried fruits of fish berry or Levant berry (*Anamirta cocculus*) are also a traditional medicine for treating parasitic worms and malaria.**

BITTER POISON

Many poisons are bitter, but the name picrotoxin, meaning 'bitter poison', was given to a mixture of toxic sesquiterpene lactones, picrotoxinin and picrotin. The fish berry (*Anamirta cocculus*) in the moonseed family (Menispermaceae), a climbing liana from Southeast Asia and parts of India, produces bunches of smooth fruits that contain picrotoxin. There are few recorded deaths from fish berry, but those that are known have resulted from preparations made from the plant. The toxins are soluble in water, and in India the crushed fruits were thrown into water to incapacitate fish. In 1890, it was reported that a 12-year-old boy had died after drinking such a poison. The fruits, often called by their pharmacopoeal drug name *Cocculus indicus* (which is also a botanical synonym), have additionally been used to treat head lice, but this practice has been abandoned as picrotoxin can be absorbed through the skin and deaths have occurred.

Picrotoxin is not only present in fish berry fruits; it is also found in Jamaica walnut (*Picrodendron baccatum*), a tree in the family Picrodendraceaea that grows on the northern islands of the West Indies. In the Dominican Republic, the species is locally known as mata becerro, literally 'calf killer', and powdered leaves have been used to kill bedbugs and lice. *Hyaenanche globosa* is a related species in the same family, from the Western Cape province of South Africa, whose scientific genus name means 'hyena poison'. It contains hyenanchin, a sesquiterpene lactone similar to tutin (see pages 72–73). The San people powdered seeds of the plant to make an arrow poison, and the fruits and seeds are still sometimes used to prepare carcasses that are laid out as lethal traps for hyenas and jackals.

Sheep and elephant killer

PLANT:
Coriaria species (e.g. *C. arborea* Linds., *C. myrtifolia* L., *C. japonica* A.Gray)

COMMON NAMES:
tanner's bush, redoul, tutu, ma sang

FAMILY:
tanner's bush or tanner's brush (Coriariaceae)

TYPE OF TOXIN:
sesquiterpene lactones of the picrotoxane type (tutin, coriamyrtin, hyenanchin, plus others)

SYMPTOMS OF POISONING IN HUMANS:
CIRCULATORY: a fast heartbeat
NEUROLOGICAL: altered mental state, tremors, convulsions, respiratory arrest
DIGESTIVE: diarrhoea, vomiting

The family Coriariaceae contains only one genus, *Coriaria*, which includes about 10–15 species of plant ranging from shrubby herbs to small trees, with opposite, simple leaves and small flowers that are most commonly arranged in a raceme. The fruit is an aggregate of achenes, which become enclosed by the petals, these developing into a fleshy, berry-like structure as they mature. The entire plant – except for the fleshy petals – is poisonous and contains convulsive sesquiterpenes, including tutin, which is why ingestion of the fruits is usually discouraged.

New Zealand is a centre of diversity for *Coriaria* species, where they are collectively known as tutu. The Maori have long known about the toxicity of these plants, but value the carefully strained juice of the fleshy petals. Even if it is hard to believe today, the first two sheep that arrived in New Zealand succumbed to tutu poisoning within days after being set ashore in 1773, and the early European settlers are said to have lost up to 75 per cent of their livestock before awareness of the plants became widespread.

The potency of tutin and its related toxins are evident from two New Zealand news stories involving poisoned elephants. In Otago in 1869, an elephant in a visiting exhibition supposedly died within three hours of eating tutu, and a second event occurred in Ohakune on the North Island in 1956, when circus elephant Mollie met the same fate.

BELOW **Redoul or myrtle-leaved coriaria (*Coriaria myrtifolia*) is the only European species in this genus, and is native to countries around the western Mediterranean. Its striking fruit have poisoned people in Spain, France and Morocco due to the presence of coriamyrtin.**

Today, Mollie's skull can be viewed at Auckland University's McGregor Museum. Human poisoning after ingesting tutu has become very rare, but in 2012 three visitors to the Kahurangi National Park in South Island had to seek hospital care after eating large amounts of a sweet-tasting fruit. Two of them developed seizures and required medical treatment, but all were discharged the following day.

POTENTIALLY CONVULSIVE HONEY

There are occasional outbreaks of secondary exposure to tutin through honey. This is not, however, due to presence of the compounds in the flower nectar used by the honeybees, even though they seem insensitive to tutin. The poisonous honey is instead the result of bees collecting honeydew produced by other insects, such as the passionvine hopper (*Scolypopa australis*), that feed on tutu. Insects feeding on plant sap ingest excessive amounts of sugars, and to regulate this they excrete a sweet solution called honeydew. In periods of drought, such honeydew is an alternative sugar source for bees in the production of honey. Analysis of toxic honey has shown that the majority of tutin is present as glycosides. This is probably a resistance mechanism that has evolved in the passionvine hopper, as the free tutin present in the plant is usually an effective deterrent to insects. The picrotoxane-type sesquiterpenes have been shown to affect the Malpighian tubule system present in most insects, which is comparable to the kidneys in mammals. By affecting this system, tutin actually increases its own concentration in the insect, making it more likely that the insect will die of the poison. Interestingly, all aphids, which are major sap-sucking plant pests related to the plant hoppers, lack a Malpighian tubule system.

ABOVE **Tree tutu (*Coriaria arborea*), on the South Island of New Zealand, bearing long, hanging heads of immature fruit. Tutu honeydew honey has caused several poisonings, most recently in 2008.**

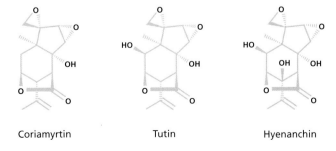

Coriamyrtin   Tutin   Hyenanchin

LEFT **The three compounds coriamyrtin, tutin and hyenanchin are very similar to picrotoxinin, and are an example of how similar compounds can evolve in distantly related plants.**

APIACEAE

Members of the carrot family are usually easily recognized when in flower, as the flowers are produced in a head, called an umbel. This gave rise to the family's classical name, Umbelliferae, meaning 'bearing umbels'. It is a large family of around 3,500 species, many of which are very familiar. We use several of these plants for culinary purposes: the vegetables carrot, parsnip, celery and celeriac, and the herbs coriander, parsley, cumin, caraway, angelica, fennel and aniseed, are all members of Apiaceae. But even though it is an important source of food and culinary plants, the group also contains some of the most poisonous plants in the world. Depending on their type, the toxins these plants contain can cause everything from a bad taste in the mouth or severe skin reactions, to seizures, paralysis and death.

FROM STINGING TASTE TO LETHAL SEIZURES

Some of the characteristic smell of carrots, and many other species of Apiaceae and its relatives, is due to a peculiar group of compounds called polyacetylenic alcohols, or polyynes. These volatile compounds are capable of strong interactions with the biochemistry of herbivores.

BELOW **An umbel-shaped flowerhead of hemlock water dropwort (***Oenanthe crocata***), with stalks arising from a single point, resembling an umbrella.**

Most polyynes – like falcarinol, present in carrots and parsnips – are rather innocuous to humans if eaten, but can cause rashes and contact allergies after repeated skin exposures. This is a particular problem for farmers and grazing animals, as the compounds are enriched in the stalks and leaves of the plants. Sometimes, the presence of polyynes becomes obvious when you eat carrots or parsnips, recognizable as a stinging or burning sensation in the taste, especially when the vegetable is unpeeled or the top parts are eaten. Polyynes can also reach higher concentrations if the growing conditions are less than optimal, or if the plants are affected by fungi or pests.

Among the most poisonous Apiaceae compounds (thankfully present in just a few species) is a small subgroup of these polyynes, the most well known of which include cicutoxin, from cowbane (*Cicuta virosa*), and oenanthotoxin, from hemlock water dropwort (*Oenanthe crocata*) (see pages 76–77). Although they are very structurally similar to the other polyynes, these particular compounds can cause lethal seizures, even after only very small amounts of the plants are ingested. They also lack the stinging taste sensation that prevents large amounts of the allergenic compounds being eaten. There are several examples of death occurring after cowbane or hemlock water dropwort have been mistaken for edible Apiaceae roots.

WHEN THE SUN ADDS TO THE SUFFERING

While polyynes can cause dermatitis, some Apiaceae species have developed chemical compounds that produce far more serious skin reactions. These compounds belong to a class called furanocoumarins, and many of them are of a type named psoralens. Their effects can be unexpected as they exert their toxic effect primarily when combined with exposure to ultraviolet radiation, such as in sunshine. The painful blisters may not develop until several days after contact, when the affected skin is exposed to the sun. Well known for causing such blistering is the giant hogweed (*Heracleum mantegazzianum*) (see pages 128–129) and its close relatives, which in parts of the world have become invasive species after escaping from cultivation. These plants can be a magnificent and stunning feature in gardens, but can spread uncontrollably if allowed to set seed.

LEFT **Psoralens in the sap of cow parsley (*Anthriscus sylvestris*) have caused rashes on exposed skin after string trimming (strimming) verges in early summer.**

THE LETHAL CUP OF SOCRATES

A number of plants in the carrot family are extremely poisonous, one being hemlock (*Conium maculatum*) (see pages 100–101). This species contains the alkaloid coniine, a neurotoxin that causes muscular paralysis and death by asphyxiation. It was used as a means of execution in ancient Greece, famously in the case of Socrates, who was sentenced to death in 399 BC. Based on contemporary accounts of the symptoms exhibited by the philosopher, the cup of poison he drank contained hemlock.

Coniine

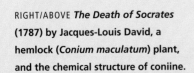

RIGHT/ABOVE ***The Death of Socrates*** **(1787) by Jacques-Louis David, a hemlock (*Conium maculatum*) plant, and the chemical structure of coniine.**

Water hemlocks – polyynes

The general name 'water hemlock' is used for species from two different genera in the carrot family, *Cicuta* and *Oenanthe*, both of which grow in damp or wet conditions, often along watercourses. These plants contain highly poisonous polyynes (unsaturated, long-chain alcohols) in all their parts, and their similarity in appearance to edible species such as the common vegetables celery or parsnip has caused many cases of human poisoning, which have included deaths.

Killer carrots

PLANT:
Cicuta virosa L. and *Oenanthe crocata* L.

COMMON NAMES:
C. virosa – cowbane, water hemlock, Mackenzie's water hemlock, northern water hemlock; *O. crocata* – hemlock water dropwort, dead man's fingers

FAMILY:
carrot (Apiaceae)

TYPE OF TOXIN:
polyynes (cicutoxin; oenanthotoxin)

SYMPTOMS OF POISONING IN HUMANS:
CIRCULATORY: a fast heart rate
NEUROLOGICAL: dilated pupils, coma, seizures
DIGESTIVE: nausea, vomiting, diarrhoea
OTHER: respiratory impairment, rapid muscle breakdown (rhabdomyolysis), renal failure

The genus *Cicuta* contains only four species, all highly poisonous and native to North America, with cowbane also found in Europe and northern Asia. In contrast, the genus *Oenanthe* is ten times larger, with 40 species ranging from North America, Africa and Europe to Australasia. Although *Oenanthe* contains several highly poisonous species, at least one, the Java water dropwort (*O. javanica*), is edible and used as a vegetable.

BELOW Hemlock water dropwort (*Oenanthe crocata*), a native of western Europe and Morocco, produces fresh green growth in spring, but it is its roots that more usually cause poisoning in humans and livestock.

BELOW The polyyne alcohols cicutoxin and oenanthotoxin produce muscle spasms very similar to those caused by strychnine. If the hydroxyl groups are altered, allergenic compounds are produced instead of convulsive poisons.

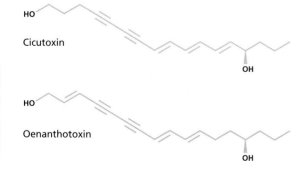

What not to do on spring holiday

Two families spending a spring holiday together ran out of things to do, and after searching the Internet for possible activities, they decided to try out a recipe for 'famine food' porridge made from the roots of the common reed (*Phragmites australis*). The roots were supposedly easy to identify by their hollow core and division into successive chambers. By the water's edge of the nearby lake the group found several roots fitting the description, floating among the reeds. The roots were peeled, cooked and made into a mash with cream and butter. Although the group cautiously tasted only small amounts, they started to feel nauseous within 30 minutes. Three young girls had enjoyed the mash and eaten two to three tablespoons. They developed vomiting and seizures, necessitating hospital care, but luckily recovered completely. The roots were identified as cowbane, and analysis of the mash showed the presence of cicutoxin and derivatives in concentrations (by weight) ten times higher than in raw roots. Unfortunately for the family, the butter and cream they added during the cooking process probably protected these toxins from destruction.

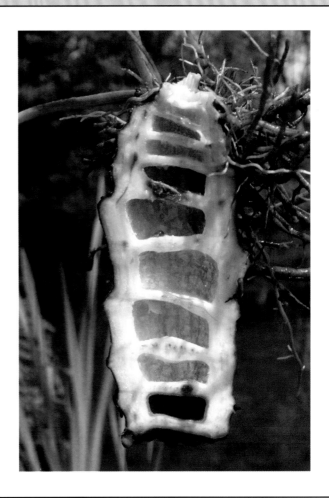

RIGHT **Cowbane (*Cicuta virosa*) is widely distributed in the northern hemisphere, where the open chambers in its roots enable it to float at the edge of waterbodies.**

THE ROOT CAUSE

The roots are the most poisonous part of the plants and exude a yellowish liquid when cut, which darkens on exposure to air. There is an easy way to differentiate between the two genera: *Cicuta* species have a tap root with several hollow compartments or chambers, while *Oenanthe* roots are solid and often consist of a bundle of several storage tubers for each plant. The roots can detach easily, allowing the plants to spread during spring flooding. The fruits are also adapted to dispersal by water, as they have a spongy outer layer that initially keeps them afloat.

Livestock have died from eating cowbane and hemlock water dropwort, with the roots of the plants being a particular hazard if they become exposed, such as during clearance of drainage systems on agricultural land. The roots are at their most toxic in winter, and this toxicity remains high but is also passed to the newly developing stems and leaves in the spring. As little else is available to eat at this time, these parts of the plant are also a cause of livestock poisonings.

GIVING A FORCED SMILE

From ancient Greek writings attributed to Homer comes the phrase *risus sardonicus*, which has become established in the medical literature and denotes the expression caused by contraction of the facial muscles, for example in cases of advanced tetanus or poisonings by toxins such as strychnine (see pages 66–67). The effects are raised eyebrows, large eyes and an open-mouthed grin. According to the classical text, the term is named after the island Sardinia in the Mediterranean Sea, where hemlock water dropwort is common. It has been stated that colonizing Phoenicians gave a stupefying potion to criminals before they were executed that produced a smile on their face. As strychnine was unknown in Europe until the early modern period, it has been suggested that the potion used was based on hemlock water dropwort due to the similar symptoms caused by oenanthotoxin.

From bees to brains – grayanotoxins

Some rhododendrons and other members of the heather family – such as bog rosemary (*Andromeda* spp.), *Kalmia* species (including mountain laurel, *K. latifolia*) and pieris (*Pieris* spp.) – produce a particular type of diterpene, the grayanotoxins. The main grayanotoxins in common rhododendron (*Rhododendron ponticum*) are grayanotoxin I (also known as andromedotoxin) and grayanotoxin III. They are toxic to both vertebrate and invertebrate animals, inhibiting sodium channels in the cell membranes, and so preventing inactivation of the nerves, primarily in the parasympathetic nervous system. Livestock deaths from eating plants containing grayanotoxins are quite common in many countries. The susceptibility of sheep to *K. angustifolia* has even given rise to some of its common names: sheep laurel and, to remove any doubt, sheep kill laurel.

Risky rhododendron

PLANT:
Rhododendron ponticum L.
COMMON NAMES:
common rhododendron, pontic rhododendron, komar
FAMILY:
heather (Ericaceae)
TYPE OF TOXIN:
grayanotoxin-type diterpenes (grayanotoxin I and III)

SYMPTOMS OF POISONING IN HUMANS:
CIRCULATORY: low blood pressure, slow heart rate, life-threatening arrhythmias
NEUROLOGICAL: sweating, blurred vision, dizziness, weakness, altered mental state
DIGESTIVE: nausea, vomiting

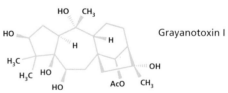

Grayanotoxin I

ABOVE **Grayanotoxin I is one of the poisonous diterpenes produced by plants in the heather family (Ericaceae), and can accumulate in harmful amounts in so-called 'mad honey'.**

LEFT **Sheep laurel (*Kalmia angustifolia*) is cultivated for its attractive flowers in temperate regions of the world, but in its native eastern United States it has poisoned livestock, including sheep.**

Common rhododendron is native to Turkey around the Black Sea, and is also found on the Iberian Peninsula. Since its introduction to areas of northwest Europe in the late 1800s, the species has become widely distributed and is invasive in habitats such as heathland.

Bee careful!

Xenophon, an Athenian military commander, was probably the first person to describe 'mad honey' poisoning, in his report of a Greek campaign against the Persian King Artaxerxes II in 401 BC. During that campaign, Xenophon's army was incapacitated for several days by mad honey as they travelled through the Black Sea region of Turkey. In a translation of his words, 'The effect upon the soldiers who tasted the honeycombs was that they all went for the nonce quite off their heads, and suffered from vomiting and diarrhoea, with a total inability to stand steady on their legs.'

Given the prevalence today of common rhododendron in some areas of northwest Europe, it might be expected that honey from these regions could contain grayanotoxins and give rise to 'mad honey' poisoning. However, this does not appear to be the case. It is only recently that research into bee behaviour has found the reason why. A study in the United Kingdom looked at the effects of grayanotoxins on three different bees: the European dark honeybee (*Apis mellifera mellifera*), the buff-tailed bumblebee (*Bombus terrestris audax*) and a solitary mining bee (*Andrena carantonica*). They found that grayanotoxin I did not affect the survival or behaviour of the bumblebee, and was sub-lethal to the solitary bee, which was deterred from feeding and showed malaise behaviour. In contrast, feeding grayanotoxin I to the honeybees was lethal, killing them within six hours. In fact, observation has indicated that honeybees do not forage on common rhododendron where it has been introduced. As it is only the honeybee that produces commercial crops of honey for human consumption, nectar from common rhododendron will not make a significant contribution to honey gathered in the United Kingdom.

By contrast, the honeybees in the eastern part of the common rhododendron's range (subspecies *Apis mellifera caucasia* and *A. m. anatoliaca*) readily forage on this plant and so produce 'mad honey'. Even so, not all Turkish honey contains grayanotoxins as the bees also gather nectar from other plants. The local beekeepers can tell when the grayanotoxins are present, as the honey causes a sharp, burning sensation in the back of the throat. This bitter honey is used medicinally but has also resulted in numerous poisonings, although no fatalities have been attributed to it in the last 30 years.

ABOVE **The nectar of common rhododendron (*Rhododendron ponticum*) is a useful food source for the buff-tailed bumblebee (*Bombus terrestris audax*), which can drink it without being poisoned.**

The identity of the red flower

A grandmother in Hong Kong was encouraged to use a traditional remedy of 'red flowers' to treat her sick two-month-old grandson. He was coughing and had previously been treated for bronchiolitis. She found a plant matching the description growing close to her home, and used it to prepare a decoction that she then administered to the infant. After he began vomiting and developed twitching in his arms and legs, she took the boy to the emergency department. Besides the twitching, he had pinpoint pupils, a slow heart rate and low blood pressure, all symptoms of activation of the parasympathetic nervous system. After successful symptomatic treatment, the boy recovered fully after two days. Analysis of his urine, and the plant material, showed the presence of grayanotoxin I, explaining the poisoning and identifying the plant as a species of *Rhododendron*. This was most likely a misidentification of the 'red flower', whose intentional identity has yet to be revealed.

The original witch's brew – tropane alkaloids

Members of the potato family that contain tropane alkaloids are among the most notorious poisonous plants. They have captured the imagination through stories of murder, fascinating poisonings and flying witches, and by their very names.

Tropane alkaloids found in plants from the potato family are usually derived from tropine. The most significant are atropine (a mixture of (S)-hyoscyamine and (R)-hyoscyamine) and scopolamine (also known as hyoscine). Which tropane alkaloids are present and their relative amounts varies between species, plant parts and the time of year, and will affect which symptoms are the most marked.

Tropane alkaloids have anticholinergic (especially antimuscarinic) effects, and the symptoms they produce fit the mnemonic taught to medical students for this class of compound:

- mad as a hatter – altered mental state
- hot as a hare – dry skin and fever
- red as a beet – flushed skin
- blind as a bat – dilated pupils
- dry as a bone – dry mucous membranes

In addition, these alkaloids will raise the blood pressure and inhibit nerve signals regulating the heart rhythm, causing the heart to start beating faster. Some of these effects have made the plants useful as medicines, explain why such uses can have unwelcome side effects and let you know what to expect if one of these plants is eaten intentionally or accidentally.

BELOW ***Atropa bella-donna*** is named after Atropos, the Greek fate who cut the thread of life, and *bella donna*, literally 'beautiful lady', probably from its use as eye drops by Italian ladies.

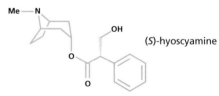

LEFT **The joined rings sharing a nitrogen atom in tropine is typical of tropane alkaloids.**

ABOVE **Atropine is made up of two forms of hyoscyamine, with the (S)-form most active. The similar scopolamine has an epoxid ring, which makes it more easily absorbed by the brain.**

The deadliest nightshade

PLANT:
Atropa bella-donna L.
COMMON NAMES:
deadly nightshade, belladonna, dwale
FAMILY:
potato (Solanaceae)
TYPE OF TOXIN:
tropane alkaloids (atropine (mixture of (*R*)-hyoscyamine and (*S*)-hyoscyamine))

SYMPTOMS OF POISONING IN HUMANS:
CIRCULATORY: palpitations, tachycardia (rapid pulse), raised blood pressure
NEUROLOGICAL: confusion, hallucinations, fever, tremors, myoclonic seizures
DIGESTIVE: dry mouth, ileus (constipation through slowed movement)
OTHER: mydriasis (dilated pupils), flushed skin, urinary retention, inhibition of sweating

We must, of course, start with deadly nightshade (*Atropa bella-donna*), an herbaceous perennial that dies back to a rootstock every year but can grow into a substantial plant during the summer. Attractive, juicy black berries follow its solitary bell-shaped purplish flowers, and conspicuous green sepals form a star at the base. Deadly nightshade grows naturally in Europe, west Asia and north Africa. In northern Europe it is particularly found on chalky soils and close to former abbeys and monasteries, where it was grown as a medicinal plant during the Middle Ages.

Belladonna is the name most often used for drugs made from the roots and flowering plants of the species, which continue to have medicinal applications for conditions such as colic of the intestines. They are included in pharmacopoeias from both within the species' natural range, such as Europe, and in countries where it is not native, including China, Japan and the Americas. The alkaloid atropine, in the form of its sulfate salt, is still used today as a pharmaceutical drug to dilate pupils in preparation for eye surgery (see page 41).

Witches' sabbat

Deadly nightshade is just one of the tropane alkaloid-containing plants that are inextricably linked to tales of witchcraft. Together with mandrakes (*Mandragora* spp.; see page 82) and henbane (*Hyoscyamus* spp.), it is said to have been an ingredient of an ointment used by witches to give them the sensation of flying, and is why witches are often depicted on broomsticks.

One of the most recent deaths from deadly nightshade, of which there are actually very few, was of a modern-day witch who went by the name of Robert Cochrane. He lived in Slough, United Kingdom, where he started a coven known as the Clan of Tubal Cain, based on a combination of Celtic mysticism and village witchcraft philosophy (American branches are known as the '1734 tradition'). Cochrane died nine days after the eve of the summer solstice in 1966, seemingly after having ingested deadly nightshade leaves and sleeping tablets. The inquest into his death returned a verdict of suicide with deadly nightshade. In some witchcraft circles it is believed that Cochrane had appointed himself as a male sacrifice.

Poison pie

Most cases of poisoning by deadly nightshade have involved the juicy black berries, which are tempting for children and misidentified by adults as edible species such as blueberries. A typical case involved a married couple who picked some fruit when out on a walk and later cooked and ate what they thought was a blackberry and blueberry pie. Their neighbours noticed that they had not opened their curtains the next day, and could get no response when they knocked on the door. The police were called and found the house in disarray and the couple in need of medical attention. The remains of a fruit pie were in the kitchen and seeds were seen in the couple's vomit. The fruits were later identified as blackberries and deadly nightshade at the Royal Botanic Gardens, Kew. As this happened in February, the identification was initially confusing, as neither of these plants were fruiting in the wild at the time. This mystery was, however, solved once it became known that the fruit had been picked and frozen the previous summer.

BELOW **Black henbane (*Hyoscyamus niger*) is native to much of Eurasia and is now widely distributed in temperate regions.**

The shrieking mandrakes

PLANT:
Mandragora officinarum L. and *M. autumnalis* Bertol.

COMMON NAMES:
M. officinarum – mandrake, devil's apples; *M. autumnalis* – mandrake, autumn mandrake

FAMILY:
potato (Solanaceae)

TYPE OF TOXIN:
tropane alkaloids (hyoscine, atropine, cuscohygrine (mandragorine))

SYMPTOMS OF POISONING IN HUMANS:
as for *Atropa bella-donna* (see page 81)

Native to countries around the Mediterranean, mandrakes are stemless plants consisting of a rosette of leaves arising from a long, branched tap root. Bell-shaped flowers nestle at the base of the leaves and are followed by round fruit.

The numerous superstitions prompted by the shape of the roots and their medicinal and toxic effects have inspired many authors. In Act IV of William Shakespeare's *Romeo and Juliet*, Juliet is worried that after taking the potion intended to send her into a death-like sleep, she will wake in a tomb to hear the spirits of the dead, which she likens to 'shrieks like mandrakes torn out of the earth, That living mortals hearing them run mad'. Mandrake really came to international popular attention, however, through its starring role in *Harry Potter and the Chamber of Secrets*, where it was an essential ingredient of the reviving potion for those who had been petrified after seeing a basilisk. Students had to wear ear protectors when repotting the young plants, as the cry of their human-like roots would otherwise knock them out for a couple of hours.

ABOVE **Mandrake (*Mandragora* spp.) with human-like roots in the *Naples Dioscurides*, a seventh-century Greek herbal.**

BELOW LEFT **Autumn mandrake (*Mandragora autumnalis*) is the more widespread of two European members of the genus. It has lilac flowers and a low-growing rosette of rough-textured leaves.**

Soporific deception

Mandrake has a long history of medicinal use; it was, for example, one of the ingredients of the 'soporific sponge', a popular medieval anaesthetic (see page 201). Its soporific qualities were known far earlier than this, however, and used with fatal consequences. In his *Strategemata* (AD 84–96), Roman senator Sextus Julius Frontinus wrote of Maharbal, a Carthaginian officer under Hannibal, who in around 200 BC was sent to fight 'rebellious Africans'. Knowing that the enemy liked wine, he mixed a large quantity with mandrake. After an insignificant skirmish, Maharbal withdrew and that night pretended to flee from his camp, leaving behind the drugged wine, which was found and drunk by his opponents. Maharbal is then said to have returned and either taken them prisoner or slaughtered them while they lay stretched out as if dead.

Angel's trumpets and devil's apples

Datura is the mostly widely distributed genus of tropane alkaloid-containing plants, being native to North and Central America, parts of Africa, much of Europe and most of Asia, and cultivated as an ornamental and occurring as a weed elsewhere. Common names for these plants include devil's apple, thornapple and jimsonweed. They are usually annuals with sturdy, branching stems and long trumpet-shaped flowers that are held either horizontally or somewhat erect.

Angel's trumpets are perennial shrubs or small trees from South America. They are included within *Datura* by some, but are generally considered to belong to a separate genus, *Brugmansia*. Thought to be extinct in the wild, these attractive plants are cultivated both in their countries of origin and also as tender ornamentals elsewhere. Their large trumpet-shaped flowers usually hang down and can be strongly perfumed at night, suggesting that they are pollinated by bats or moths. The exception is *B. sanguinea*, which has scentless red and yellow flowers that are pollinated by hummingbirds.

The name jimsonweed (a contraction of 'Jamestown weed') is based on an incident of 1676 in Jamestown, Virginia, where British soldiers sent to quash a rebellion made a boiled dish using a weed. The description published in 1705 by historian Robert Beverley, Jr. (c. 1667–1722) relates the symptoms as 'a very pleasant comedy', with one soldier blowing a feather up into the air while another threw straws

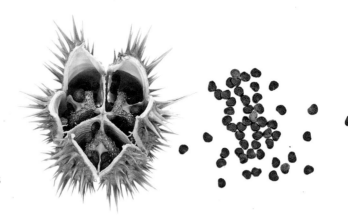

ABOVE **Jimsonweed (*Datura stramonium*) fruit capsule splitting open to reveal seeds in six chambers rather than the more usual four. Some of the black kidney-shaped seeds have been removed.**

at it, a third sitting naked in a corner smiling and making faces toward passers-by, and yet another running around kissing and patting his companions. The symptoms lasted for 11 days.

Periodically, stories like this revive experimentation, by adolescents and adults, with the potential hallucinogenic properties of angel's trumpets and other tropane-containing plants, usually with serious consequences.

BELOW **Jimsonweed (*Datura stramonium*), with toothed leaves, erect trumpet-shaped white flower and immature fruit capsule. Native to the southern United States and Central America, it is now an almost global weed.**

Herbal hallucinations – isoquinoline alkaloids

The amaryllis family (Amaryllidaceae) is now divided into three subfamilies following the inclusion of the African lilies (*Agapanthus* spp.; subfamily Agapanthoideae) and the genera previously treated as the onion family (subfamily Allioideae). It is, however, only the species included in the older, narrower circumscription (now subfamily Amaryllidoideae) that are considered poisonous to humans. They contain a unique group of more than 500 alkaloids derived from a common precursor (norbelladine). Here, we discuss the poison bulb (*Boophone disticha*); other examples are given later in the book (see pages 142–143 and 212–213).

From ox killer to medicine men

PLANT:
Boophone disticha (L.f.) Herb.
COMMON NAMES:
poison bulb, bushman poison bulb, Cape poison bulb, candelabra flower, gifbol, incotho, seeroglelie (literally 'sore-eye flower')
FAMILY:
amaryllis (Amaryllidaceae)
TYPE OF TOXIN:
Amaryllidaceae alkaloids (buphanidrine and others)

SYMPTOMS OF POISONING IN HUMANS:
CIRCULATORY: high blood pressure, tachycardia (quick heart rate)
NEUROLOGICAL: hallucinations, altered mental state, muscle rigidity, unconsciousness
DIGESTIVE: nausea, vomiting
OTHER: raised temperature, laboured breathing

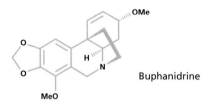

ABOVE **Buphanidrine is an isoquinoline alkaloid of a structural type characteristic of the amaryllis family (Amaryllidaceae). Its hallucinogenic activity may explain its use in traditional medicine.**

The poison bulb is an iconic plant distributed from east Africa to Namibia and the Western Cape province of South Africa. The plant's large bulb produces an inflorescence of red or pink flowers in spring. The greyish-green leaves, which grow in a wide fan shape, are large – up to 45 cm (18 in) long and 5 cm (2 in) wide – and often have wavy margins. When the seeds have matured, the dry seedhead breaks off and the seeds disperse as wind blows it across the ground.

An old arrow poison

Rock paintings in southern Africa depict plants that have been interpreted as *Boophone*, and a 2,000-year-old mummy discovered in the Kouga Mountains in South Africa was found covered, to a large extent, with a thick layer of poison bulb leaves. Together with analysis of more modern artefacts, such as poison arrows collected in 1806 showing the

LEFT **In spring, a short-stalked head of fragrant flowers emerges from the partially exposed bulb of the poison bulb plant (*Boophone disticha*). The leaves appear after flowering and grow into a distinctive fan shape.**

presence of Amaryllidaceae alkaloids, it is evident that poison bulb has been an important cultural plant in this part of the world for a long time. Although the scientific name of the genus *Boophone* is derived from Greek and literally means 'ox killer', poison arrows tipped with *Boophone* were most probably used only for hunting smaller animals.

ABOVE **Another name for the poison bulb (*Boophone disticha*) is the fireball lily, which perfectly describes the heads of pink flowers, each with six slender tepals. Wild populations of this southern African plant are declining due to the harvesting of large quantities for traditional medicine.**

MEDICINE AND MAGICAL POTION

The poison bulb is a common medicinal plant in South Africa, and has been used both among native people and more recent immigrants. Among the Sotho and Xhosa, it is an important part of the rite of passage from boy to man, including its use in dressings for circumcision wounds, and Dutch settlers reportedly treated hysteria and insomnia by sleeping on mattresses stuffed with bulb leaves. Decoctions of poison bulb are drunk as emetics and purgatives, as aphrodisiacs and memory enhancers, and as treatments for stress-related ailments and mental illness. However, ingesting poison bulb, or the use of decoctions as enemas, is associated with an apparently large risk of poisoning: historically, as many as 30 deaths a year have been ascribed to poison bulb use as a traditional medicine or means of suicide.

In many traditional systems of medicine, the cause of disease is intertwined with the spirit world. As poison bulb can cause inebriation and hallucinations, it is used for these properties by traditional healers and witch doctors in South Africa to contact the spirits, and diagnose and treat patients or even curse them. Among the Manyika tribe of eastern Zimbabwe, plants are cultivated outside dwellings to bring good luck, keep away the spirits of the recently deceased and ward off evil dreams.

Manslaughter under the influence

A man who believed that he had been put under an evil spell was given 150 ml (5 fl oz) of a decoction by a traditional healer to induce a trance state, under which the healer would be able to uncover the culprit responsible for the bewitchment. After a short time, the patient started to hallucinate, became agitated and thought he was under attack. To protect himself, he took out a gun and started firing randomly, wounding several people and killing one. After his arrest, the decoction was analysed and shown to contain compounds present in poison bulb, including buphanidrine and other alkaloids, together with eugenol, an aromatic, volatile oil.

Stimulating brain chemistry – *beta*-carboline alkaloids

For thousands of years, humans have used a number of plants that stimulate brain chemistry to alter their mental state, including the peyote cactus (see box opposite and page 65) and any of the 50-plus plant species from various families where tryptamines such as *N,N*-dimethyltryptamine (DMT) have been found. Yopo (see page 65), a member of the legume family (Fabaceae), is one such plant; its seeds are used in the Orinoco Basin of Colombia and Venezuela to make a hallucinogenic snuff. There are reports of it being used as early as 1496 in the West Indies, where it was called cohoba. The seeds of harmala or Syrian rue (*Peganum harmala*) in the family Nitrariaceae, from which harmine, a *beta*-carboline indole alkaloid, was first isolated, are used as ritual incense in north Africa and Arabian countries. It is, however, a drink, ayahuasca, which is usually a mixture of plants containing *beta*-carboline alkaloids and tryptamines, that now has a global reach.

BELOW Twisted stem of the ayahuasca vine (*Banisteriopsis caapi*) in the Peruvian Amazon. Wild plants are becoming more scarce, but the species is easy to propagate and often planted near urban areas.

Ayahuasca – 'vine of the soul'

PLANT:
Banisteriopsis caapi (Spruce ex Griseb.) Morton (syn. *B. inebrians* Morton)
COMMON NAMES:
ayahuasca vine, caapi, yagé
FAMILY:
Barbados cherry or acerola (Malpighiaceae)

TYPE OF TOXIN:
beta-carboline alkaloids (harmine and others)
SYMPTOMS OF POISONING IN HUMANS:
NEUROLOGICAL: dizziness, dilated pupils, aggressive behaviour, altered visual perception
DIGESTIVE: nausea, salivation, vomiting, diarrhoea

The ayahuasca vine, also known as caapi or yagé, among numerous other names, is a member of the Barbados cherry or acerola family (Malpighiaceae). It grows in the moist tropical climate of South America, including the western regions of the Amazon Basin, Colombia and Peru. The large vine has attractive pink flowers, but it is usually the bark, and sometimes also the leaves and roots, that are used by indigenous peoples to make a drink, known as ayahuasca. Traditionally, this drink is taken only by men for ceremonial purposes, divination or healing.

Ceremonies involving ayahuasca last for many hours, with small amounts of the liquid being drunk an hour or so apart. The effects depend on many factors, including the setting, the intended purpose and any control exercised by the shaman.

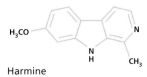

Harmine

N,N-dimethyltryptamine

ABOVE **Indole alkaloids may interfere with neurotransmitters in several ways. Harmine is a *beta*-carboline that inhibits the breakdown of *N,N*-dimethyltryptamine (DMT), enabling the latter compound to be absorbed into the brain, where it causes hallucinations.**

ABOVE **Ayahuasca drink being prepared in Pucallpa, Peru, by cooking sections of the stem or bark of the ayahuasca vine (*Banisteriopsis caapi*) and chacruna (*Psychotria viridis*) leaves in water.**

Ayahuasca is prepared with the addition of other plants, the most common being chakruna, or chacruna (*Psychotria viridis*), in the coffee family. It is claimed that different forms of ayahuasca vine influence the experiences of the shaman, with at least one classification being based on the animal that he will be transformed into: a jaguar, anaconda or goshawk. The decoction induces visionary states of consciousness, but usually also side effects that include nausea, vomiting, diarrhoea and shivering.

The active compounds in ayahuasca vine are *beta*-carboline alkaloids, including harmine. They prevent the breakdown of the DMT present in chakruna leaves when the two plants are combined, and it is the DMT that causes a modified state of consciousness characterized by introspection, visions and enhanced emotions.

Set and setting – the safety of hallucinogens

Ayahuasca use has now moved out of its original forest setting and into more populated rural areas in the region, where it is sometimes combined with Catholic customs to form the basis of ayahuasca churches. Since the 1960s, ayahuasca tourism has become popular, and the drink has also been exported to the West. Outside its traditional setting and control, there have been a few instances where serious harm has occurred. At least two young men have died after participating in ayahuasca ceremonies, one in southwest England and one in Maryland, United States. However, in both cases post-mortem examinations found the presence of additional psychoactive drugs such as opium poppy (*Papaver somniferum*; see pages 200–201), cannabis (*Cannabis* spp.), and tryptamines presumed to be synthetic. In contrast to these serious events, a significant number of studies have suggested that ayahuasca preparations are reasonably safe and have potential benefits in areas such as anxiety, depression and substance abuse. However, the large variations in composition of ayahuasca makes these results very unreliable.

Ancient rituals

Among evidence of human occupation in caves along the Rio Grande in Texas, United States, archaeologists found two peyote 'buttons', as the dried tops of the small peyote cactus are commonly known. They were placed in the collections of the Witte Museum in San Antonio. Radiocarbon dating has given these buttons a mean age of 5,700 years, and chemical analyses detected the presence of mescaline in both samples. This makes them the oldest known plant material that still contains a major bioactive compound. Other items collected with the peyote (which are also known as peyotl and mescal buttons) are similar to objects used today by Native Americans during peyote ceremonies, and provide evidence that humans were collecting and using the cactus as early as 3780–3660 BC.

Seeing and believing – ergot alkaloids

This book may be about killer plants, but fungi could not be kept out completely as there are some that are so closely associated with plants that they exert their effects through them. In fact, an increasing number of plants considered to be medicinal or toxic are being found to contain active compounds actually produced by fungi that live on or inside the plant or are induced by their presence. The most famous and well studied of these plant–fungal pairs are those between plants in the bindweed and grass families (Convolvulaceae and Poaceae, respectively), and fungi from the family Clavicipitaceae that produce what are commonly known as ergot alkaloids, from which the synthetic hallucinogen lysergic acid diethylamide (LSD) can be derived.

Heavenly blue

PLANT:
Ipomoea tricolor Cav.
COMMON NAMES:
morning glory
FAMILY:
bindweed or morning glory (Convolvulaceae)
TYPE OF TOXIN:
ergot (ergoline) alkaloids (ergometrine, ergine (lysergic acid amide, or LSA))

SYMPTOMS OF POISONING IN HUMANS:
CIRCULATORY: high blood pressure
NEUROLOGICAL: dizziness, weakness, hallucinations, convulsions
DIGESTIVE: nausea, vomiting

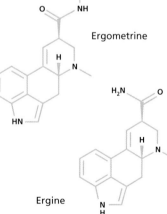

Ergometrine

Ergine

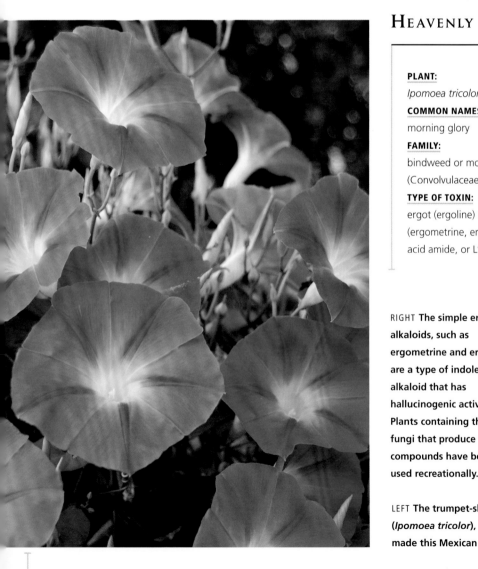

RIGHT The simple ergot alkaloids, such as ergometrine and ergine, are a type of indole alkaloid that has hallucinogenic activity. Plants containing the fungi that produce the compounds have been used recreationally.

LEFT The trumpet-shaped blue flowers of the morning glory (*Ipomoea tricolor*), related to the sweet potato (*I. batatas*), have made this Mexican climber a popular tender garden plant.

The seeds of several species of morning glory, including the commonly cultivated ornamental *Ipomoea tricolor* (sometimes incorrectly referred to as *I. violacea*), along with ololiuqui or Christmas vine (*Turbina corymbosa*, syn. *Rivea corymbosa*) and Hawaiian baby woodrose (*Argyreia nervosa*), contain significant concentrations of ergot alkaloids that have been synthesized by fungi with which they have a symbiotic relationship. Other commonly grown relatives, for example *I. purpurea*, have not been found to contain such alkaloids.

Ololiuqui and morning glory were both used widely to enable divination during healing practices and religious ceremonies in Mexico before the Spanish conquest in the sixteenth century. Both are twining climbers with heart-shaped leaves, trumpet-shaped flowers and dry fruit. Ololiuqui is a woody perennial, bearing many-flowered heads of white flowers with greenish stripes, which are followed by fruits that each contain a single round brown seed. In contrast, morning glory is an annual, bearing heads of three to four flowers that are shades of blue, red or white, and are followed by fruit that contain several angular black seeds.

In modern times, the seeds of morning glory, ololiuqui and Hawaiian baby woodrose (which can easily be purchased) have been used by young people in search of psychoactive effects. As with all such experimentation, there is a risk of overdose, complications arising from concomitant use of other drugs and injury while under the influence. In one incident in Denmark, for example, a young man lost his life by jumping from a window and falling four floors after eating Hawaiian baby woodrose seeds and smoking cannabis.

Poisonous pumpernickel

Bread made from the seeds of rye grass (*Secale cereale*) in the grass family has been a staple of the human diet in large areas of Europe, Scandinavia and west Asia for millennia. Although not poisonous themselves, rye grass and other cereals can host the ergot fungus (*Claviceps purpurea*), which produces highly toxic ergot alkaloids, principally ergometrine and ergotamine. Throughout history, many people have been poisoned and killed by eating contaminated rye grass products. Evidence that humans have been exposed to these toxins includes the presence of traces of ergot in the stomachs of the Scandinavian Iron Age bog bodies known as Grauballe man and Tollund man (dating from the third to fourth centuries BC).

The dark sclerotium of the ergot fungus takes the place of a grass seed and is harvested and processed along with the grain. This gives bread made from the contaminated flour a very unpleasant taste, but the poor people who relied on it could not afford to be fussy. Consumption of such bread gave rise to two distinct forms of poisoning: a gangrenous form (see box), and a convulsive form. In the convulsive form of ergotism, numbness and pins and needles in the limbs are followed by twitching facial muscles. In more severe cases, agitation and epileptic-like spasms lead to multiple convulsions and eventually death.

St Anthony's fire

Ergotamine and ergometrine are powerful vasoconstrictors, which in high doses can even completely stop the flow of blood to the fingers and toes. If this lack of circulation is prolonged, the tissues will start to die and become gangrenous. The Hospital Brothers of St Anthony was founded in the late eleventh century to treat the sick, especially those afflicted by the painful, burning symptoms of ergot poisoning, and hence the condition became known as 'St Anthony's fire'. The monks established a strong reputation for treating these patients, possibly by using flour from uninfected grains when baking bread. However, with the identification of the fungal cause of ergotism in 1676 and the instigation of preventative measures, the number of cases dwindled and the order was subsumed within the Knights Hospitaller. Today, ergot alkaloids have a limited use in the treatment of migraines, and a semi-synthetic derivative, bromocryptine, can relieve certain symptoms of Parkinson's disease.

RIGHT **Spores of the ergot fungus (*Claviceps purpurea*) infect the developing ovules in the head of rye grass (*Secale cereale*), where they grow into a hooked, dark purple to black structure called a sclerotium.**

CHAPTER 5

MORE THAN A WEAKNESS OF THE KNEES

Besides the effects on nerve signalling within the brain, there are several other ways a poison can affect peripheral muscle function. This chapter explores examples of poisons that work on muscle-related nerve signalling outside the brain, and often outside the central nervous system (CNS). The system of final executive muscle control can be inhibited or stimulated, effects that humans have utilized, for example, when hunting with arrow poisons and in the execution of criminals.

ATTACKS ON MUSCLES: MECHANISMS OF ACTION

We usually feel that we are in control of the muscle functions that enable all our movements, but in the previous chapter we encountered plants that could disrupt this control by inducing seizures through muscle contraction. In this chapter we look at plants that deprive us of our muscle control by causing paralysis through muscle relaxation.

THE NEUROMUSCULAR JUNCTION

Between the motor neurons responsible for initiating movement and the muscle that is actually affected there is a small gap or synapse, called the neuromuscular junction or motor end-plate (see page 39). The neuron releases the neurotransmitter acetylcholine, which crosses the synapse and activates nicotinic acetylcholine receptors on the muscle. This initiates a movement of calcium ions inside the muscle, which leads to its contraction and corresponding movement of the body.

Several plant compounds can interact at this interface between the nervous system and the skeletal muscle, causing paralyzing muscle relaxation by blocking the nicotinic receptors and eliminating the signal transmission. The most drastic plant

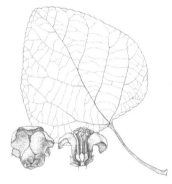

LEFT Leaf and male flowers of pareira brava (*Chondrodendron tomentosum*), the source of one type of curare with the active compound *d*-tubocurarine.

poisons that stun animals in this way are the curares, which actually consist of two very distinct types of compounds (see pages 94–97). Curares are ideal for use as arrow poisons because only negligible amounts are absorbed from the human intestinal tract, meaning that small animals killed in this way are usually perfectly safe to eat. There are also compounds that can stimulate the activity of the nicotinic receptors at the neuromuscular junction. These receptors are named after the nicotine present in tobacco (*Nicotiana tabacum*), which was found to stimulate them, and severe poisonings from nicotine and other piperidine alkaloids cause muscle cramps, among several other symptoms (see pages 98–101).

LEFT Tobacco plants (*Nicotiana tabacum*) tied to stakes at harvest prior to being transported to the curing barns. Originally from the Americas, tobacco is the world's largest non-food crop.

LEFT **Golden chain or golden rain (*Laburnum anagyroides*)** is named for its hanging heads of bright yellow flowers. The plant contains quinolizidine alkaloids, and even the flowers have caused poisoning when eaten in fritters.

INSECTICIDES

In mammals, the nicotinic receptors can be found both in the CNS and peripheral nervous system (PNS), but in insects they are confined to the CNS. This widespread presence in animals is probably one evolutionary reason why there are so many plants that contain compounds that interfere with these receptors and act as feeding deterrents or even lethal poisons. Nicotine itself has been used as both a pesticide and the model for a series of synthetic insecticides called neonicotinoids, which due to their preferential binding to insect and not to mammalian receptors are considered to be safer options than nicotine itself. Although neonicotinoids are selective for insects, they do not, however, discriminate between useful species such as honeybees and pests like plant hoppers or aphids. It is suspected that neonicotinoids may contribute to colony collapse disorder in honeybees, and consequently they are now heavily regulated in Europe and North America.

NOT ONLY IN MUSCLES

Different types of nicotinic receptors are also present in the CNS, both in the spinal cord and in the brain, and these are activated by the quinolizidine alkaloids present in many members of the legume family (Fabaceae) (see pages 104–105). In the spinal cord, these alkaloids activate neurons in both the sympathetic and parasympathetic nervous systems, whose effects in the body usually counteract each other. These nicotinic receptors usually transmit their signals through nerves that mediate their effects at the target by affecting muscarinic acetylcholine receptors. They therefore influence a number of different bodily functions, as described in the discussion on tropane alkaloids on pages 80–83. Stimulation of the parasympathetic nervous system leads, among other things, to salivation, lacrimation (tear production), increased production of digestive enzymes and increased movement of the intestines.

The nicotinic receptors in the brain are involved in several cognitive effects and also in the addictive effects of tobacco. The stimulation of these receptors increases alertness and the ability to focus, and has a positive effect on short-term memory. It also leads to increased activity in the reward system of the brain, where nicotinic receptor activation makes it more sensitive to dopamine and prolongs the sensation of pleasure. This is the reason why the habit of smoking (and other uses of tobacco) is so hard to break.

BELOW **Honeybee collecting nectar from a lavender flower.** Some insecticides, including neonicotinoids, can poison beneficial insects as well as pests.

Paralyzing plants – curares

Effective arrow poisons do not all have the same mechanisms of activity. In Chapter 3 we looked at heart-stopping poisons, and in this chapter we will focus on paralyzing toxins, of which the curares are the classic example of muscle-relaxant plant poisons. Europeans were introduced to the paralyzing arrow poisons of the Amazonian rainforest by reports from Spanish conquistadors in the early sixteenth century. The name curare is considered to be derived from *ourari*, a word used for the poison by a companion of Sir Walter Raleigh (c. 1554–1618) during his visit to Venezuela in 1594. Other forms of the word used by Europeans include *ourara*, *urali*, *woorari* and *wourali*, which are possibly a distortion of local words meaning 'bird' or 'to kill'. Due to the wars between the colonial powers, the nature of curare was not investigated in detail until the beginning of the nineteenth century.

On the expected death of a donkey

By 1812, it had been demonstrated that death from curare occurs as a result of paralysis of the skeletal muscles, including those of the diaphragm, which are responsible for breathing. In a paper published that year by the physiologist and sometime royal physician Sir Benjamin Brodie (1783–1862), it was further shown that small animals could survive an injection of curare if they received assistance with breathing, in the form of bellows to inflate the lungs, for the duration of the paralysis. This was later demonstrated in a display using three donkeys injected with curare in front of an audience. The first animal died after an injection in the shoulder. The second donkey had a tourniquet applied to its foreleg and was injected below the constriction; it survived until the tourniquet was released and the poison was distributed by the bloodstream. The third donkey seemed to succumb to the injection but was kept alive through assisted ventilation. After the surviving donkey revived, it lived out its time under the name Wouralia in recognition of the poison it had survived.

Curare composition

When the chemical composition of curare was eventually investigated, it was evident that there were several different types of the poison, containing different groups of alkaloids. The poison could often be identified by the container in which it was traditionally carried: a tube, calabash or pot. These three groups roughly indicate both geographical as well as botanical origin. Tube curare, or bamboo curare (the original tubes were made out of bamboo), is found in the western Amazon rainforest of northern Peru and eastern Ecuador, and is often based on species from the moonseed family (Menispermaceae). Calabash curare comes from northern South America east of the Orinoco River and north of the Amazon River and its Río Negro branch, and is usually derived from *Strychnos* species (Loganiaceae). Between these two areas is a region where curare is stored in clay pots (pot curare) and contains alkaloids from both the moonseed and Loganiaceae families.

LEFT **Pareira brava (*Chondrodendron tomentosum*) is a South American liana whose ability to paralyze has been exploited as an arrow poison, as well as for use in surgery during the twentieth century.**

Tube curares

PLANT:
Chondrodendron tomentosum Ruiz & Pav.
COMMON NAMES:
pareira brava, curare vine
FAMILY:
moonseed (Menispermaceae)

TYPE OF TOXIN:
bisbenzylisoquinoline alkaloids (*d*-tubocurarine)
SYMPTOMS OF POISONING IN HUMANS:
MUSCLE: rapid onset of paralysis

d-tubocurarine

LEFT The alkaloid *d*-tubocurarine, present in curare, is a muscle-relaxant bisbenzylisoquinoline that blocks the signal transmission between nerves and muscles.

BELOW Huaorani hunters in the Amazonian region of Ecuador use blow darts tipped with curare that has been ritually prepared by a shaman as a sign of respect to the hunted prey.

Tube curares are derived from several species in the moonseed family, of which only one has received wider recognition: pareira brava (*Chondrodendron tomentosum*). This tropical liana grows in Central and South America. The leaves are large, 10–20 cm (4–8 in) long and almost as wide, somewhat heart-shaped and have a white velvety underside. The plants are dioecious, forming racemes of small greenish-white male or female flowers. The female plants bear small, fleshy, pear-shaped bluish to black fruits with a bitter-sweet taste that are considered edible.

The genera used for tube curares all contain dimeric alkaloids of the benzylisoquinoline type. Denoted by the prefix 'bis-', dimeric alkaloids are formed from two subunits of connecting alkaloid precursors. Bisbenzylisoquinoline alkaloids are present in several groups of plants, but it seems that the curare activity is restricted to members of Menispermaceae. The most important curare alkaloid, *d*-tubocurarine, has been reliably isolated only from pareira brava, although other species contain bisbenzylisoquinolines with similar activity, such as bebeerine (curine) and isochondrodendrine. Tubocurarine has been developed into the pharmaceutical drug atracurium, which is included on the *WHO Model List of Essential Medicines* as a peripheral muscle relaxant used in surgical procedures.

Calabash curares

PLANT:
Strychnos toxifera
R.H.Schomb. ex Lindl.
COMMON NAMES:
devildoer, curare
FAMILY:
Loganiaceae

TYPE OF TOXIN:
monoterpene indole alkaloids (toxiferine I)
SYMPTOMS OF POISONING IN HUMANS:
MUSCLE: rapid onset of paralysis

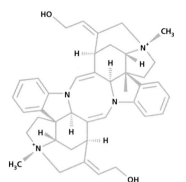

Toxiferine I

LEFT **The alkaloid toxiferine I shares structural features with the seizure-causing toxin strychnine and is isolated from the same plant genus, but it acts as a muscle relaxant instead.**

The *Strychnos* genus contains nearly 200 species widely distributed in the tropics. They are often important for use in rituals or as medicine, with different physiological effects on animals depending on which major alkaloid is present. At least a dozen South American *Strychnos* species have been identified as components of calabash curare. The most studied of these is devildoer (*S. toxifera*), a vine with simple opposite leaves, and a branched inflorescence of hairy tubular flowers that are followed by berry-like fruits. It contains several alkaloids that can be considered dimeric forms of the same precursor as strychnine, the toxin in the seizure-inducing nux vomica (*S. nux-vomica*) from Asia (see page 66). In contrast to strychnine, the *Strychnos* curares act on the neuromuscular junction, causing paralysis. The main alkaloid involved is toxiferine I, which has been developed into a muscle-relaxing drug called alcuronium. The alkaloids are primarily extracted from the bark of devildoer but seem to be present in most of the plant. *Strychnos* curares are not only found in some species from the Amazon, but have also been isolated from a few arrow poisons of central Africa and Southeast Asia, where poisons based on cardioactive steroids or convulsants such as strychnine are more commonly used.

BELOW **Devildoer (*Strychnos toxifera*) is a vigorous climbing vine from Colombia and Venezuela, with heads of tubular flowers that are hairy on their outer surface.**

The case of Dr X

In 1975, *New York Times* journalist Myron Farber (b. 1938) was handed an anonymous letter by his editor, telling the story of an unnamed chief surgeon who had murdered as many as 40 patients at his hospital, again unnamed. During the investigation, Farber came across the case of Dr Mario Jascalevich (1927–1984), who in 1966 had been investigated for nine suspicious deaths at his hospital after the discovery of 18 mostly empty vials of tubocurarine in his locker. That investigation had been closed as Jascalevich had explained that he had used the paralytic compound in experiments on dogs at Seton Hall University. With the publication of Farber's article in 1976, the investigation was reopened. In the subsequent trial, in which Jascalevich was accused of five murders using tubocurarine, the jury eventually acquitted him, relying on the defence witnesses' refutation of the testimonies for the prosecution that claimed to have found tubocurarine in tissues from the five exhumed victims from 1966. Though medically trained murderers have been convicted of using synthetic muscle relaxants in other cases, this is the only known instance where the defendant was accused of using the natural alkaloid.

Bean and Gone

PLANT:
Erythrina americana Mill.,
E. berteroana Urb., and others
COMMON NAMES:
Erythrina spp. – coral tree;
E. americana – colorín, tzompantli
FAMILY:
legume (Fabaceae)

TYPE OF TOXIN:
benzylisoquinoline alkaloids (erythrinan alkaloids, including erysodine)
SYMPTOMS OF POISONING IN HUMANS:
MUSCLES: rapid onset of paralysis
NEUROLOGICAL: sedative, possibly hallucinogenic

ABOVE **Bright red coral tree (*Erythrina* spp.) seeds, known as colorines or lucky beans, are used decoratively in jewellery and also medicinally in carefully controlled doses.**

The curare arrow poisons are not active if ingested because the intestinal absorption of these alkaloids is usually very low. There is, however, a plant genus that contains an orally active muscle relaxant: *Erythrina*, in the legume family. The active compounds in *Erythrina* are a distinct group of modified benzylisoquinolines called erythrinan alkaloids, of which erysodine is an example with curare-like effects. The genus grows worldwide in tropical and subtropical regions, with around half of its approximately 130 species being present in Central and South America. They are usually trees, 6–30 m (20–100 ft) tall, with trifoliate leaves and racemose inflorescences. The red flowers are usually pollinated by birds, including long-billed hummingbirds in the case of some specialized American species, whose flowers consequently have a distinct bean-pod-like appearance. In parts of Central America, the flowers are eaten and allegedly possess mild hypnotic effects.

Erythrina beans (seeds), called colorines, are often bright red and are incorporated into decorative necklaces and amulets. They are also used in traditional medicines, where their toxicity is recognized and the dosage is usually constrained to a quarter or a maximum of half a bean. It is reported that the indigenous people of Michoacán, Mexico, used an extract of colorines to exact revenge on their foes.

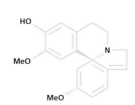

Erysodine

LEFT **Erysodine and similar isoquinoline alkaloids can cause paralysis even after ingestion, in contrast to the classical curare alkaloids, which are not absorbed from the intestines.**

RIGHT **The colorin tree (*Erythrina americana*) from Mexico is spectacular in spring, when the heads of scarlet flowers, each up to 10 cm (4 in) in length, develop before the leaves open.**

Stimulant or poison – piperidine alkaloids

Many plants that today are recognized as poisonous were originally introduced to Europe as socially acceptable recreational drugs rather than for medicinal purposes. Notable examples include strychnine and cocaine, which were incorporated into tonics and medicinal spirits as pick-me-ups before their flaws of high toxicity and development of drug dependence, respectively, became apparent. There is, however, one poisonous plant that has survived as a common stimulant and even conquered the world despite the adverse effects and recognized risks of its use: tobacco.

From ritual to habit

PLANT:
Nicotiana tabacum L., N. rustica L., N. glauca Graham

COMMON NAMES:
N. tabacum – tobacco;
N. rustica – Aztec tobacco;
N. glauca – tree tobacco

FAMILY:
potato (Solanaceae)

TYPE OF TOXIN:
piperidine alkaloids (nicotine, anabasine)

SYMPTOMS OF POISONING IN HUMANS:

CIRCULATORY: initially tachycardia (rapid heart rate) and hypertension, in severe cases followed by bradycardia (slow heart rate), hypotension and arrhythmias

NEUROLOGICAL: initially confusion, dizziness, small pupils and tremors, in severe cases followed by lethargy, coma, large pupils and muscle paralysis (including respiratory muscles)

DIGESTIVE: nausea, vomiting, salivation, diarrhoea

Nicotine

Anabasine

ABOVE Nicotine and anabasine are the major alkaloids in species of *Nicotiana*, and are responsible for both the recreational use of the plants and in the production of insecticides.

LEFT Tobacco plants (*Nicotiana tabacum*) can grow to 1–3 m (3–10 ft) tall and produce 10–20 leaves. The pink flowers are often removed on crop plants to encourage more leaves to grow.

The tobacco genus (*Nicotiana* spp.), in the potato family (Solanaceae; see pages 138–139), contains 76 species, with the greatest diversity in the Americas and Australia. They are mostly herbaceous, a few species being shrub-like, with simple leaves and tubular flowers that are often pollinated by hawk moths, hummingbirds or even bats. The plants produce a number of related alkaloids, the most well known being nicotine, but its analogue anabasine has caused several lethal poisonings in humans. These piperidine alkaloids act through stimulation of acetylcholine receptors, with high concentrations leading to an overstimulation that actually inhibits the normal functions of these receptors.

Although the oldest use of nicotine-containing plants is probably by the Aborigines of Australia, it was from the Americas that the practice of smoking the leaves was brought to Europe in the sixteenth century. Species of tobacco were used as medicinal plants by several Amazonian tribes, and there is archaeological evidence of tobacco cultivation in what is now Mexico from around 1400–1000 BC. However, it is the North American Indians who have become the embodiment of the original tobacco smoker. Among the multitude of tribes, the tobacco plant was often regarded as a supernatural gift to humans, and the smoke was considered to carry prayers to the spirit world.

Tobacco smoking became a more mundane activity after its introduction to Europe. The first automated cigarette-manufacturing machine was patented by inventor James Bonsack (1859–1924) in 1881, making it possible to produce 200 cigarettes a minute. This made them cheap, and by 1940 the number of cigarettes smoked per capita in the United States was 2,000 a year, with this figure exceeding 4,000 per annum 20 years later. With the weight of scientific evidence now supporting a causative correlation between smoking and lung cancer, the number of smokers has been decreasing in the western world. Even so, more than 7 million tonnes of cultivated tobacco were still produced in 2014.

ABOVE **Coyote tobacco (*Nicotiana attenuata*), from the western United States, including the Sierra Nevada, has branched heads of tubular white flowers that are pollinated by hawk moths. Its fruit are small capsules that dry to brown when ripe.**

From insecticide to smoking prevention

It is thought that species of tobacco produce piperidine alkaloids to protect them from herbivorous insects. Their effectiveness is evident by the large number of commercial insecticides based on nicotine available throughout the world (see page 93). Insects are more sensitive than mammals to nicotine, due to differences in the distribution of receptors, and experience more pronounced CNS effects. Despite the presence of these defensive alkaloids, several moths feed on tobacco plants and some species of tobacco have developed additional mechanisms to minimize harm. The night-flowering coyote tobacco (*Nicotiana attenuata*) is usually pollinated by hawk moths, but it is also a food plant for their caterpillars. When attacked by such larvae, the plant changes pollination tactic, decreasing the size of its flowers and opening them in the morning instead of at night. This attracts hummingbirds instead of moths, and possibly decreases the risk of attracting further egg-laying moths.

As human poisoning with tobacco or tobacco plant material usually leads to early vomiting, products containing the pure alkaloid, which does not illicit this response, constitute a much greater hazard for lethal poisonings. Nicotine-replacement products are a fairly recent development, used to increase the success rate during smoking cessation. Many of these are based on the principle of replacing the addictive nicotine in tobacco with the pure compound, and they come in a variety of forms, from gums, tablets and skin patches, to liquids used in electronic cigarettes and other vaporizers. Eliminating the actual smoking is expected to reduce the risk of lung cancer, making tobacco less of a killer plant.

THE TRUE HEMLOCK

PLANT:
Conium maculatum L.

COMMON NAMES:
hemlock, poison hemlock, carrot fern, poison parsley

FAMILY:
carrot (Apiaceae)

TYPE OF TOXIN:
piperidine alkaloids (coniine, and similar)

SYMPTOMS OF POISONING IN HUMANS:
CIRCULATORY: rapid heart rate
NEUROLOGICAL: large pupils, trembling, low body temperature, muscle paralysis (finally including respiratory muscles)
DIGESTIVE: nausea, vomiting, salivation, diarrhoea
MUSCLES: rhabdomyolysis (muscle breakdown)

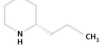

Coniine

ABOVE **Although it has a simple structure, the alkaloid coniine is a powerful muscle relaxant, eventually causing death by respiratory paralysis. It is the most toxic of the piperidine alkaloids found in hemlock (***Conium maculatum***).**

Although the name hemlock seems to have been applied to several poisonous plants, *Conium maculatum* is the species that should be considered the 'true' hemlock. In Chapter 4 we encountered the water hemlocks, also from the carrot family (Apiaceae), which cause potentially lethal seizures by disrupting the CNS. Here we look at the opposite effects, where the alkaloid coniine causes muscle paralysis, leading to respiratory arrest in severe intoxications.

Flowering plants of the biennial species hemlock, a native of Europe, north Africa and temperate Asia, can reach 2.5 m (8 ft) in height. The slightly ridged, hollow stems are spotted or mottled with purple flecks, and the leaves are fern-like and often have an odour resembling mouse urine. The entire plant contains several highly poisonous piperidine alkaloids, of which coniine is the most toxic. They act in a similar manner to nicotine, although they also produce muscle paralysis similar to that caused by curares (see pages 94–96).

The toxicity of hemlock has been known for a very long time, and during the Hellenistic period (323–31 BC) the species was employed as a poison in the execution of criminals (see box, page 75). Hemlock has been used as an externally applied analgesic and internally as a sedative and antispasmodic, and even as an antidote against strychnine poisoning. The use of hemlock leaves was included in the British Pharmaceutical Codex until 1934, and up until the First World War almost 21 tonnes (23 tons) of seeds and dried leaves were imported into the United States each year. As it was used for asthma, epilepsy and whooping cough, it is not hard to imagine that this 'treatment' actually contributed to the risks of these diseases.

LEFT **Hemlock (***Conium maculatum***) is a striking plant growing in damp ground such as road verges. Its umbels of white flowers have small bracts (leaf-like structures) at the base of the main umbel and even smaller bracteoles at the base of the partial umbels.**

ABOVE It is unwise to eat fool's parsley (*Aethusa cynapium*), whose small plants grow to 50 cm (1.5 ft). It is most easily identified when in flower or fruiting, as the partial umbels have long bracteoles.

Know what you eat

As with many modern poisonings, the risk of hemlock ingestion correlates to the ability of the forager to identify the plant. The carrot family contains a number of root vegetables and culinary herbs, and several severe and even fatal poisonings with hemlock have occurred when it has been misidentified as parsnip (*Pastinaca sativa*), fennel (*Foeniculum vulgare*), parsley (*Petroselinum crispum*) and carrots (*Daucus carota*). All these plants belong to the same family, and their superficial similarities can explain the accidental ingestion of hemlock, emphasizing that knowledge about the purple-spotted stem and the unpleasant smell of this plant can be life-saving. There is, in addition, another species in the carrot family that may be as lethal as hemlock. Fool's parsley (*Aethusa cynapium*) also contains coniine, and is recognized by the long bracteoles below the partial umbels.

The hand of God

The Book of Numbers in the Old Testament tells us that during their travels after their exodus from Egypt, the Israelites became tired of eating only manna. Their complaints angered God, and although he gave them meat in the form of quail, he also warned that they would loathe this meat later. As the people ate the birds, they were struck by a plague and many died before the march towards the Promised Land was resumed. It has been speculated that this story is a description of what has become known as coturnism, a poisoning that occurs in the Mediterranean region during the seasonal migration of European common quails (*Coturnix coturnix*), which are trapped and eaten. The symptoms include nausea, vomiting, muscle weakness and rhabdomyolysis, potentially leading to renal failure. As the quail migration coincides with the ripening of hemlock seeds, and the symptoms of coturnism mirror those of hemlock poisoning, it is likely that the condition is a result of secondary poisoning by coniine.

BELOW Illustration of the Old Testament story of the Israelites gathering quails (*Coturnix coturnix*) in a place they named Kibroth Hattaavah during their exodus from Egypt.

FABACEAE

The legume family, Fabaceae, is one of the largest plant families, with some 19,500 species in more than 750 genera. At times its members have been treated in the smaller families Caesalpiniaceae, Mimosaceae and Papilionaceae, based on differences in flower form. Probably the more familiar name for this family, however, is the now outdated Leguminosae, which has the same origin as the common name legume. Whatever it is called, this is one of the most important plant families for human survival, producing a variety of edible pulses, peas, beans and lentils that are a rich source of protein.

FROM STINGING TASTE TO LETHAL SEIZURES

Being such a large and diverse family, it is no wonder the legumes include several poisonous species, many of which are covered in this book. They include the hallucinogenic yopo bean (*Anadenanthera peregrina*; see pages 65 and 86), curare-like coral tree beans (*Erythrina* spp.; see page 97), widespread cytisine-containing species (see pages 104–105), purgative Alexandrian senna (*Senna alexandrina*; see page 144), blood-thinning mouldy sweet clover (*Melilotus* spp.; see pages 170–171), and plants that produce aberrant amino acids to protect against herbivory, such as the grass pea (*Lathyrus sativus*; see pages 186–187).

RECOGNIZABLE FLOWERS AND FRUITS

As mentioned above, there are three basic forms of Fabaceae flowers: the tight clusters of small, regular flowers in the mimosoid species, such as mimosas or acacias (for example, *Senegalia* spp., *Vachellia* spp. and *Acacia* spp.); the open flowers with spreading petals of peacock flowers (*Caesalpinia* spp.), flamboyants (*Delonix* spp.) and sennas (*Senna* spp. and *Cassia* spp.); and the classic pea flower, with the petals forming the prominent keel and sail parts, including peanut (*Arachis hypogaea*).

Regardless of the flower form, however, almost all species produce easily recognized fruits in the form of legumes (or pods) containing one or several seeds. The seeds of the garden pea (*Pisum sativum*) and broad bean (*Vicia faba*), and the entire pods, including the seeds, of green beans (*Phaseolus vulgaris*) and runner beans (*Phaseolus coccineus*), are probably familiar to us all as vegetables. With other legumes we sometimes eat the seeds once they and the pod have dried, such as kidney beans (a form of *Phaseolus vulgaris*) and peanuts. More unusually, the seeds may be encased in a fleshy pulp, as in the tamarind (*Tamarindus indica*).

BELOW **The flamboyant or flame tree (*Delonix regia*) has open flowers that are bilaterally symmetrical. Endemic to Madagascar, it is now widely grown in tropical and subtropical regions.**

A PASSING NOTE FROM AN ENTOMOLOGIST

Maria Sibylla Merian (1647–1717) was a German-born botanical artist and, during her lifetime, was also considered to be a leading entomologist. She is most famous for documenting butterfly metamorphosis and for her detailed botanical illustrations. In 1699, Merian was awarded a travelling stipend to the Dutch colony of Surinam, where she not only sketched plant and animal life, but also became interested in the situation of the local people and slaves working on the Dutch plantations. On such an occasion she learnt that it was common to use roots of the peacock flower (*Caesalpinia pulcherrima*) as an abortifacient to prevent children being born into slavery, or even to commit suicide. After returning to the Netherlands, Merian published this information in her 1705 book *Metamorphosis Insectorum Surinamensium*. Although the local knowledge of this use was again reported during the late twentieth century, the underlying compound responsible for the effect has yet to be described.

ABOVE **Peacock flower (*Caesalpinia pulcherrima*) illustration from *Metamorphosis Insectorum Surinamensium* (1705) by Maria Sibylla Merian.**

CHEMICAL DIVERSITY GUARANTEES SUCCESS

Members of the legume family grow practically all over the world in any type of biome except the driest deserts and the northernmost Arctic climates. This might be due to the fact that many species collaborate with nitrogen-fixing bacteria, giving them an extra edge in nutrient-poor environments, but it may also be due to the unique diversity of the small molecular compounds they contain. In no other family have chemical investigations yielded so many different types of toxic substances. The examples above produce several types of alkaloids, polyketides and aberrant amino acids, but the family also contains species that produce fluorinated acetic acid (the poison pea genus, *Gastrolobium*; see page 181), as well as toxic proteins (see pages 146–147). There are even some species that produce cardenolide-type poisons (for example, in the genus *Coronilla*) or cyanogenic glycosides (some clovers, *Trifolium* spp.).

LEFT **Following pollination, the stem bearing the peanut (*Arachis hypogaea*) flower pushes the developing seedpod into the ground.**

Pods of poison – quinolizidine alkaloids

Among the many different toxins found in the legume family (see pages 102–103), the alkaloids of the quinolizidine type are especially common. These alkaloids have activities similar to those of nicotine (see pages 98–99), but can also cause anticholinergic effects such as a dry mouth and difficulty with urinating (see pages 80–83). Hundreds of different quinolizidine alkaloids have been described, with cytisine being the most studied. This alkaloid was originally isolated from scotch broom (*Cytisus scoparius*), from which it got its name, but is present in large amounts in other 'broom' species, which are actually placed in several different genera, including *Genista*, *Spartium*, *Ulex* and *Retama*. In addition, cytisine is found in golden chain (*Laburnum anagyroides*) and alpine laburnum (*L. alpinum*), kowhai and related species (*Sophora* spp.), and mescal beans (*Dermatophyllum* spp.).

Beautiful but hazardous beans

PLANT:
Dermatophyllum secundiflorum (Ortega) Gandhi & Reveal (syn. *Sophora secundiflora* (Ortega) Lag. ex DC.)
COMMON NAMES:
Texas mescal bean
FAMILY:
legume (Fabaceae)

TYPE OF TOXIN:
quinolizidine alkaloids (e.g. cytisine)
SYMPTOMS OF POISONING IN HUMANS:
CIRCULATORY: tachycardia and hypertension
NEUROLOGICAL: confusion, dizziness, tremors
DIGESTIVE: nausea, vomiting, diarrhoea
OTHER: urine retention

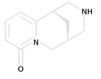

Cytisine

ABOVE **Cytisine is a widespread quinolizidine alkaloid in many legumes and has similar effects to nicotine. It has been used in the development of nicotine-replacement therapies.**

RIGHT **Flowering plant of the Texas mescal bean (*Dermatophyllum secundiflorum*), also known as the Texas mountain laurel. This evergreen shrub or small tree is native to Mexico and Texas and New Mexico in the United States.**

The Texas mescal bean (*Dermatophyllum secundiflorum*) is a small shrub or tree with pinnate leaves, fragrant purple flowers that form dense clusters and hard, beaded pods containing large orange-red seeds – the mescal beans. A brew made from mescal beans was used in religious ceremonies by several American Indian tribes of the North American plains; its propensity to cause vomiting was considered cleansing, driving out evil, and the intoxicating effects gave spiritual insight. There are several similarities between the use of mescal beans and peyote cactus (*Lophophora williamsii*) in Native American rituals (see page 87), and it has even been suggested that peyote was used instead of mescal beans to avoid the risks associated with the latter. Despite the fact that the hallucinogen in peyote, called mescaline, was named after the ritual use of mescal beans, the compound is not present in these legumes. This confusion of names and origins of intoxicating plants also extends to mezcal, the alcoholic spirit derived from different species of agave (*Agave* spp.), where the name is derived from *mexcalli*, a Nahuatl word meaning 'oven-cooked agave'.

ABOVE **The bright yellow flowers of golden chain (*Laburnum anagyroides*; see page 93) are followed by small pods. These contain soft green seeds that harden and turn black as they ripen.**

Troublesome *turmus*

In the Mediterranean area, the seeds of the white lupin (*Lupinus albus*; see photograph) are eaten as a snack, sometimes called *turmus*. White lupins have been cultivated since the Bronze Age and today several strains exist, classified into sweet and bitter cultivars depending on the amount of alkaloids they contain. Unless the bitter seeds are soaked in water for several hours to leach away the alkaloids, especially lupinine, they may cause poisonings. For the unsuspecting physician, these poisonings can mimic strokes, as the patients are usually severely muddled and debilitated by dizziness, and may vomit and have very wide pupils. In places with large Mediterranean minorities but where lupin seeds are not usually consumed, such as Sweden and Australia, these poisonings often cause confusion among doctors, but are usually resolved within a day or two.

Chain reaction

The small golden chain tree is often considered to be a very poisonous plant and a substantial hazard for children. The species has trifoliate leaves and long, pendulous racemes of yellow flowers, which produce clusters of pods with small seeds that turn black when ripe. Although the immature green seeds and pods can be very tempting to children (for example, it was reported in 1979 that in an average summer more than 3,000 children in England and Wales are admitted to hospital after eating them), severe poisonings are extremely rare due to their vomit-inducing (emetic) activity. In modern times there have, however, been two lethal intoxications from golden chain, when two individuals ate 23 and 25 pods, respectively. Both were being treated at the time with medications that suppressed the emetic activity of the cytisine, allowing the full amount of toxin to be absorbed.

Kicking the habit

Cytisine's action as a partial activator of the nicotinic receptors in the brain has been used to help smokers fight their psychological dependence on nicotine. In eastern Europe, paper strips were soaked in cytisine and put under the tongue or lip, to be absorbed without eliciting emetic effects, but due to the risks associated with overdosing they are no longer marketed. The pharmaceutical drug varenicline, based on cytisine, is instead used today as an alternative to nicotine-replacement therapies.

CHAPTER 6

STARTING AS AN IRRITATION

Plants use a wide array of compounds that, while not considered lethal or acutely dangerous, can cause local irritation and are responsible for large amounts of suffering. This chapter looks at plants containing compounds that lead to skin reactions. Examples include the mainly irritating nettles and their far more potent relatives, as well as plants that produce potentially delayed reactions or trick our senses to detect heat sensations that do not actually exist.

SKIN REACTIONS: MECHANISMS OF ACTION

Causing immediate pain on contact is one of the ways in which plants can deter animals – including people – from damaging or eating them. We have probably all been stung by a nettle, but other plants usually need to be broken in order to release the chemicals in their sap or within their cells. Sometimes pain and inflammation are caused by sharp crystals. In other cases, chemicals released from the plant trigger sensory nerves or cause chemical reactions within the skin. Many of the substances that cause contact reactions on larger animals are also poisonous to insects or fungi that feed on the plant.

ABOVE **Nettle stem magnified to show the stinging hairs, which have a sharp, rigid tip and a swollen base that holds irritant compounds.**

RAPHIDE OXALATES

Plants in the arum family (Araceae) frequently contain needle-like calcium oxalate crystals called raphides. In species of dumb cane (*Dieffenbachia* spp.; see pages 112–113), the raphides are held in specialized cells called idioblasts; when the plant is chewed, the crystals are released and penetrate the soft tissue of the mouth. The damage caused by these crystals triggers white blood cells to release histamine, and also enables penetration by other irritant chemicals present in the plant, increasing the pain and inflammation of the soft tissue. The reaction caused by these plants is called mechanical irritant contact dermatitis due to the physical injury involved, but it shares elements of mechanism and symptoms with chemical irritant contact dermatitis.

BELOW **Microscopic view of an isolated idioblast cell from dumbcane (*Dieffenbachia seguine*), showing calcium oxalate raphides inside.**

STINGING HAIRS

Many members of the nettle family (Urticaceae), including some as large as trees, are covered in stinging hairs (see pages 110–111). The reaction they cause is called contact urticaria. Specialized hairs, or trichomes, inject irritant and pain-inducing compounds into the skin when brushed against. The tips of these rigid hairs break off, and compression of the swollen base forces the contents – the chemical cocktail – through the hollow hair and into the skin.

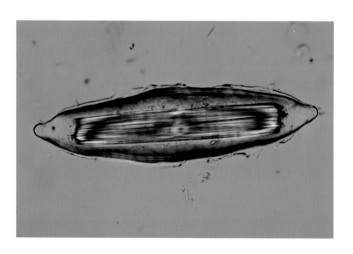

ABOVE **The milky sap of many *Euphorbia* species contains chemicals that irritate the skin and eyes on contact.**

PHOTOTOXIC

Some plants cause skin reactions only if contact with their sap occurs during or before exposure to ultraviolet (UV) radiation in sunlight. Such reactions are called phytophotodermatitis (see pages 128–129). Phototoxic plants are found in several families, including the citrus (Rutaceae), carrot (Apiaceae), legume (Fabaceae) and mulberry (Moraceae) families The substances in the plant sap, for example psoralens, are absorbed by the skin, where they can then be activated by UV radiation and react with deoxyribonucleic acid (DNA) in the skin cells, leading to cell death and extensive local inflammation.

BELOW **Common rue (*Ruta graveolens*) in the citrus family causes blistering if skin comes into contact with its sap in bright sunlight.**

CHEMICAL IRRITANTS

Contact reactions from a number of plants are caused by the presence of chemicals that irritate the skin, mouth and eyes. They achieve this either by stimulating specific receptors on sensory neurons, or by binding to protein in the skin. The chemicals that cause this chemical irritant contact dermatitis include:

- diterpene esters in the latex-like sap of many members of the spurge family (Euphorbiaceae) (see pages 116–119);
- isothiocyanates, highly pungent chemicals in 'mustard oil', produced by members of the cabbage family (Brassicaceae) when damaged (see pages 120–121);
- capsaicin, the pungent compound that makes chilli peppers 'hot' (see pages 122–123); and
- protoanemonin (see pages 126–127), a volatile substance produced by some members of the buttercup family (Ranunculaceae) when they are damaged.

ALLERGENS

Many plants are mildly allergenic and can cause a rash on sensitive individuals, called allergic contact dermatitis. Poison ivy and poison oak (*Toxicodendron* spp.), however, are in a league of their own (see pages 130–131). Their allergen, urushiol, is highly sensitizing and the rash that results from contact can last for weeks or even months. Urushiol induces an allergic reaction by binding to skin proteins. Sensitivity to the allergen is an acquired immune response, so, unlike in other forms of dermatitis, only individuals with previous exposure develop allergic effects.

LEFT **The German primula (*Primula obconica*) is an allergenic plant, although the number of allergic reactions has reduced since the breeding of allergen-free cultivars.**

Stinging hairs

Stinging hairs are found in a number of different genera in the nettle family. The stinging sensation is followed by contact urticaria, commonly known as nettle rash, which is typically a number of individual raised red patches that correspond to the points on the skin that have been penetrated by a stinging hair, a type of trichome. The rash is itchy and painful, with the severity and duration depending on how extensive the contact was and the species of nettle involved. Stinging nettles (*Urtica* spp.), for example, are found on all continents except Antarctica, but they are not all as relatively benign as the common stinging nettle (*U. dioica*), whose stinging action has actually been used as a treatment for conditions such as arthritis.

BELOW **Gympie bush (*Dendrocnide moroides*) with pink fruit, growing in the rainforest understorey in Queensland, Australia. Skin contact results in a severe prickling effect, soon followed by an intense pain.**

INJECTING CHEMICAL PAIN

PLANT:
Dendrocnide moroides (Wedd.) Chew (syn. *Laportea moroides* Wedd.)
COMMON NAMES:
gympie bush, gympie-gympie, stinger bush
FAMILY:
nettle (Urticaceae)

TYPE OF TOXIN:
biogenic amines (histamine, serotonin, acetylcholine), organic acids and possibly small peptides
SYMPTOMS OF POISONING IN HUMANS:
SKIN: itchy and painful rash

Histamine

LEFT **Histamine is active in inflammatory responses in the body, and after injection by stinging hairs it will lead to pain, swelling and blisters.**

Investigations have shown that the transient, short-lived reaction experienced after contact with the common stinging nettle is caused by a mixture of compounds, including histamine and serotonin, while organic acids enhance the pain sensation and local inflammation. As these irritant compounds are destroyed on cooking, the young growth of common stinging nettle and other related species makes a nutritious spring green. A word of caution when collecting plants, however, comes from Thomas Morong (1827–1894), an American botanist who in 1889 described a botanical trip around Buenos Aires, Argentina:

The plant that I shall be likely to remember the longest is a nettle, very common here in waste grounds, which on first sight I took to be Urtica dioica *or* U. gracilis, *and which I boldly grasped as I had so often done at home. It proved, however, to be a much more formidable species, armed with numerous spines, the sting of which made my fingers tingle for hours afterwards. It is* Urtica spathulata *[sic], which I warn all North American botanists who follow in my track to collect with gloves on their hands.*

There are tales of even more serious consequences for humans and animals that have come into contact with 'nettles' in Australia, New Zealand and elsewhere. Horses have become incapacitated for several days or even died, and affected dogs salivate profusely, gag, vomit and stagger. Humans reported that the sting remained noticeable for several days, weeks or even years, particularly in damp weather or if the area of skin became wet. There was even one human fatality in New Guinea. In these cases, often the precise species or even genus of the offending plant remained unknown. However, attempts have been made to identify some, such as a species in Timor, Indonesia, called daoun setan or devil leaf ('*Urtica urentissima* of Blume'), tree of the settlers (*Dendrocnide gigas*, syn. *U. gigas*) and gympie bush (*D. moroides*).

ABOVE **All parts of mala mujer (*Cnidoscolus angustidens*) bear stinging hairs, including the flowers and fruit.**

Stinging spurges

Although members of the spurge family are probably better known for their chemical irritant effects (see pages 116–119), several species bear stinging hairs. In the genus *Tragia*, including the branched noseburn, *Tragia ramosa* (see below), the initial sting is due to mechanical damage by a crystal at the end of the stinging hair cell, while the contents of the cell cause a secondary reaction. Little is known of the severity of effects from species of *Tragia*, prompting a request from a taxonomist for field botanists to record the intensity of the stings they experience when collecting specimens.

TREADING CAREFULLY

In tropical and subtropical America, from Kansas in the United States to Argentina, there is a genus in the spurge family, *Cnidoscolus*, the name of which is derived from the Greek words for 'nettle' and 'prickle' or 'thorn'. Species in this genus have hairs that are similar in form and function to those of nettles (*Urtica* spp.). The stinging hairs vary between species – for example, some Mexican species have hairs up to 3 mm (1/8 in) in length, while another has hairs 10–20 mm (3/8–3/4 in) long. *Cnidoscolus megacanthus* has to win the prize for deterring herbivores, however, as its stinging hairs are found in combination with enlarged persistent thorns. The common names for these plants reflect their reputation, such as tread carefully, tread softly or finger rot (*Cnidoscolus urens* ssp. *stimulosus*, syn. *C. stimulosus*), and Texas bull nettle (*C. texanus*). Mala mujer, literally 'bad woman', is used for various poisonous plants in the spurge family, such as *Euphorbia cotinifolia* and *Jatropha gossypiifolia*, as well as several species of *Cnidoscolus*, including *C. angustidens*, a plant found in Arizona and Mexico. The stinging hairs in the latter species are known to cause severe burning and itching eruptions on contact, which can be accompanied by swollen regional lymph nodes.

Crystal spikes

Calcium oxalate crystals occur widely in the plant kingdom, being found in more than 200 families. They form within plant cells and have a variety of roles: the removal of excess oxalic acid, maintaining the balance of ions/regulating intracellular pH balance, and protecting the plant by discouraging foraging. The crystals form a number of different shapes, with plants in the arum family specializing in needle-shaped raphides. The raphides cause mechanical damage to soft tissue and enable the entry of chemical irritants, resulting in severe reactions in any animal that is unwise enough to take a bite.

STRUCK DUMB

PLANT:
Dieffenbachia seguine (Jacq.) Schott (syn. *D. picta* Schott)
COMMON NAMES:
dumbcane, leopard lily
FAMILY:
arum (Araceae)
TYPE OF TOXIN:
calcium oxalate raphides, proteases (proteolytic enzymes) and possibly other chemical irritants

SYMPTOMS OF POISONING IN HUMANS:
DIGESTIVE: irritation of the mouth and throat, burning sensation; in rare, severe cases extreme pain, swelling of the mouth, tongue and throat, thickened or unintelligible speech, or loss of voice, occasionally blistering; if swallowed, swelling and erosion of the throat and gastrointestinal tract, vomiting, diarrhoea
SKIN: itching, reddening, rash
EYES: irritation, extreme pain, corneal injury

Of all the plants in the arum family that contain calcium oxalate raphides, it is species of dumbcane (*Dieffenbachia* spp.) that are the most notorious, having been used by man for barbaric purposes that are thankfully consigned to history. This genus of more than 50 species is native to South and Central America, and is grown as an ornamental in tropical and subtropical climates. A few species, particularly *D. seguine*, are common houseplants elsewhere, largely grown for their attractively patterned leaves, which are usually variegated with spots, blotches or streaks of white or cream.

The popularity of members of the arum family as houseplants in the United States resulted in more than 61,000 calls to poisons units during a ten-year period for dumbcane alone. Other less toxic plants in the family, particularly peace lily (*Spathiphyllum* spp.) and philodendron (*Philodendron* spp.), have similarly caused thousands of calls but usually result in only mild or no symptoms. Luckily, the immediate irritation and burning that results from eating dumbcane is usually sufficient to limit the amount that is chewed and encourages the unfortunate individual to spit out the plant material rather than swallow it, thereby avoiding life-threatening damage to the gastrointestinal tract. The general mildness of symptoms resulting from accidental exposure to this plant has led some doctors to question whether it is in fact a seriously poisonous plant, but occasionally catastrophic exposures are reported, confirming its reputation as a killer.

LEFT **The variegated leaves of dumbcane (*Dieffenbachia seguine*) make it a popular houseplant around the world, but care should be taken to keep it away from pets and children, who, being unaware of the noxious calcium oxalate crystals, may be tempted to take a bite.**

ABOVE **A field of taro (*Colocasia esculenta*) plants growing in Hawai'i. These robust plants can reach 0.5–2 m (1.5–6 ft) high, and have leaf blades 20–50 cm (8–20 in) long.**

Taro – an acrid arum

Despite the presence of calcium oxalate raphides and chemical irritants, the corms of several plants in the arum family are used as an edible root crop known as taro. The most important sources of taro are cultivars of *Colocasia esculenta* (syn. *C. antiquorum*) and *Xanthosoma sagittifolium*. The latter is native to South America but spread to other continents probably in the nineteenth century. By contrast, *C. esculenta*, also known as cocoyam, dasheen and eddo, is native to the region extending from India to southeast China and south to Sumatra. It is known to be an ancient crop, with records of cultivation in Egypt at least 2,000 years ago, and is now widely grown in the humid tropics. The British botanist Joseph Banks wrote in the account of his voyage with Captain Cook on HMS *Endeavour* (1768–1771) of its cultivation in Southeast Asia and Australasia, noting 'the immense sharpness of every part of this vegetable before it is dressed makes it probable that any people who have not learned the uses of it from others may remain forever ignorant of them'.

Grown mainly on a small scale for home consumption, these are important root crops in the humid tropical regions of Africa, Asia and the South Pacific. Taro is adapted to growing in wet conditions – some cultivars even thrive in flooded areas – and is generally tolerant of shade, conditions that other root crops would not survive. The numerous cultivars vary in how 'acrid' they are, and although it is mainly the corms that are eaten, the leaves and stems of the less acrid varieties of *Colocasia esculenta* are also consumed as a green vegetable. Taro corms are prepared before being eaten by prolonged baking or boiling, or anaerobic fermentation in an underground pit for several weeks. Raw or insufficiently prepared taro causes oral pain and inflammation, with more serious symptoms occurring if a large quantity is consumed.

LEFT **Taro tubers require preparation before they are eaten, but are an important source of carbohydrate, particularly in areas with high rainfall where other crops struggle to grow.**

EUPHORBIACEAE

Spurge, the common name for the *Euphorbia* genus, is derived from the French *épurger*, in reference to the purgative properties of these plants. It is also used as the common name for the Euphorbiaceae family, many of which produce purgative oils. This large family of herbaceous plants, shrubs, small to large trees and succulents contains around 6,745 species in 218 genera, including some very large genera such as *Euphorbia* and *Croton*, and some with only one or a few species, including *Hura*. It has an almost global distribution, with most diversity of species being found in tropical and subtropical regions.

STINGING HAIRS AND IRRITANT ESTERS
A few genera bear stinging hairs, such as *Tragia* and *Cnidoscolus* (see page 111). Confusingly, some species of *Cnidoscolus* have at times been placed in the genus *Jatropha*, giving the impression that this genus can also bear stinging hairs. However, *Jatropha* species do not bear stinging hairs, but exert their toxicity through the presence of highly irritant diterpene esters throughout the plant, particularly in the latex and seeds.

The diterpene esters, of which there are several types, are the most widespread of the Euphorbiaceae toxins (see pages 116–119). They have been used by man as aids to hunting – for example, according to Spanish historian Gonzalo Fernández de Oviedo y Valdés (1478–1557), the indigenous people in Venezuela used the manchineel tree (*Hippomane mancinella*; see page 118) to make 'that diabolic poison for their arrows'. The toxic properties of these compounds make them particularly potent fish poisons, and a large number of the species that contain them are used for this purpose in Africa. Among these, the pencil tree (*Euphorbia tirucalli*) is the most important, being easy to obtain and highly effective. Its active compounds inhibit the cellular respiration of fish, leading to paralysis.

LETHAL LECTINS
Some members of the spurge family contain poisonous proteins known as lectins. The lectin from the seeds of the castor oil plant (*Ricinus communis*; see pages 148–149) is ricin, one of the most poisonous substances known. Castor oil itself, which lacks the lectin, is obtained from the pressed seeds of the plant and is used traditionally as a laxative, and also in soap, detergents, paints, dyes and lubricants, among other things. Other species, such as jicamilla (*Jatropha cathartica*) and purging nut (*J. curcas*) also contain lectins (curcin), but these lack the ability to enter cells (see page 146). Oil from seeds of purging nut is used as a biofuel.

BELOW **Purging nut (*Jatropha curcas*) is a large shrub or small tree native to Mexico and south to Argentina. Its toxicity is mainly due to phorbol esters, but its seeds also contain a lectin.**

OCCUPATIONAL HAZARDS

Due to the range of compounds produced by plants in the spurge family, many are valued as sources of rubber, oils, dyes and medicines, some of which have already been mentioned. Additional examples are the rubber tree (*Hevea brasiliensis*), a native of Brazil and the Guianas, which is the main source of natural rubber, while tung oil obtained from the seeds of the tung tree (*Vernicia fordii*, syn. *Aleurites fordii*) and related plants is used to protect wood. Care is generally required when using spurges, or extracting raw materials from them, as contact with them is highly hazardous. The wood even retains its toxicity when dry, making burning it inadvisable. Due to a lack of awareness of the risks, and the lack of protective clothing used by workers and other unsafe work practices, members of this family are the cause of a significant number of occupational poisonings, particularly in developing countries.

LEFT **Rubber trees (*Hevea brasiliensis*) are now grown in tropical countries such as Thailand, where their latex is tapped for the production of natural rubber.**

EAT IF YOU DARE

Very few plants in the particularly toxic spurge family are used as food. However, despite containing the cyanogenic glycoside linamarin, cassava (*Manihot esculenta*; see pages 184–185) is the staple food for more than 700 million people in Africa, Asia and South America. Cassava has to be processed before it is safe for human consumption, but some animals are able to include plants in this family within their diet, presumably because they can tolerate or detoxify them. The gemsbok (*Oryx gazella*) of the African savannah eats the toxic Damara milkbush (*Euphorbia damarana*) during periods of drought, and other species of spurge are eaten by greater kudu (*Tragelaphus strepsiceros*) and black rhinoceros (*Diceros bicornis*). These succulent plant species provide the herbivores with water as well as nutrients.

RIGHT **A gemsbok (*Oryx gazella*) rests in the shade of a Damara milkbush (*Euphorbia damarana*) in Damaraland, Namibia.**

Irritant diterpene esters – a burning issue

The sap of many plants in the mezereum (Thymelaeaceae) and spurge families contains compounds that are chemical irritants. These compounds are diterpene esters, for which some 20 skeletal types have been isolated to date. The effects of these different skeletal types are similar, but there can be considerable differences in the severity and duration of symptoms. Important among these are the phorbol esters (with a tigliane skeleton), ingenol esters (ingenanes), jatrophanes, lathyranes and daphnanes.

The biological activities of the phorbol esters are highly structure-specific. Those with certain structures activate protein kinase C enzymes, which in turn activate metabolic processes (such as increasing gene expression and biosynthesis of proteins and DNA) in preparation for cell division and differentiation. Even at very low concentrations, they are toxic if eaten and irritant on contact. In addition, certain phorbol esters are co-carcinogenic (promote tumour growth following exposure to a carcinogen), with tetradodecanoyl phorbol acetate being the most potent of these known to date.

The most potent irritant known

Euphorbium was an irritant plant resin that is said to have been named by King Juba II of Mauretania (50 BC–AD 23) in honour of his physician Euphorbus, after whom Carl Linnaeus also named his genus *Euphorbia*. Although the original ingredients of the resin are not precisely known, it has latterly been obtained from the resin spurge (*E. resinifera*). This leafless, succulent plant is native to the Atlas Mountains in Morocco. Its main active compound, resiniferatoxin, a daphnane diterpene, was not isolated until 1975, and in its pure form has been found to be 1,000 times hotter than pure capsaicin on the Scoville scale, at 16 billion Scoville heat units (see page 123). Resiniferatoxin acts on sensory neurons, with exposure initially causing strong irritation followed by analgesia (pain relief). This ancient medicine has been investigated as a topical treatment for the relief of pain associated with diabetic polyneuropathy and post-herpetic neuralgia following shingles.

RIGHT **Resin spurge (*Euphorbia resinifera*) from Morocco** resembles a cactus, lacking leaves and with four ridges along its length that carry short spines. Plants should be handled with extreme care.

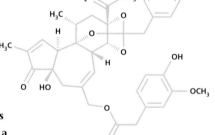

Resiniferatoxin

LEFT This diterpene compound causes intense pain through the same mechanism as capsaicin from chilli peppers.

Don't drink this milk

PLANT:
Euphorbia tirucalli L.

COMMON NAMES:
pencil tree, pencil cactus, African milkbush, avelós

FAMILY:
spurge (Euphorbiaceae)

TYPE OF TOXIN:
diterpene esters of the tigliane, ingenane and daphnane types

SYMPTOMS OF POISONING IN HUMANS:
SKIN: burning pain, inflammation, swelling, blistering
EYES: pain, keratoconjunctivitis (inflammation of the cornea and soft tissue)

Phorbol

LEFT **Phorbol is a diterpene that is present in many members of the spurge family, where its esters are responsible for causing inflammation.**

BELOW **The pencil tree (*Euphorbia tirucalli*) is a shrub or small tree with only a few inconspicuous leaves. It is used as hedging, and forms with orange or red stems are grown as ornamentals.**

The spurges or euphorbias are notorious among gardeners for the irritant effects of their milky sap. Cutting stems or simply breaking off a leaf causes this sap to ooze from the damaged plant, and it can then easily be transferred to bare skin or eyes. Depending on the species of euphorbia and the extent and duration of the contact, the inevitable effects might range from mild and transient, to severe and lasting several weeks. Luckily, one of the most well known and commonly cultivated species in the *Euphorbia* genus, poinsettia (*E. pulcherrima*), is one of the least poisonous. It is sold in its millions over winter and causes great concern when eaten by children and pets, but causes only mild irritation and discomfort.

One of the most toxic *Euphorbia* species is the pencil tree, a succulent shrub or small tree native to dry regions of India, Pakistan and Madagascar, and also east and southern Africa, where its sap is used as a fish poison (see page 114). In addition, the pencil tree is cultivated in tropical regions such as Florida and Hawai'i. If sap from the tree gets in the eye, it causes immediate burning of the eye and eyelid, tear production and sensitivity to light, followed after some 8–12 hours by blurred vision and increased pain. The outer covering of the cornea is corroded, with loss of visual clarity, and there is reddening and swelling of the soft tissue, including the eyelid. Vision is restored after one or two weeks depending on the severity of the initial reaction, and there are no long-lasting effects.

Little Apple of Death

PLANT:
Hippomane mancinella L.
COMMON NAMES:
manchineel, manzanillo, manzanilla
FAMILY:
spurge (Euphorbiaceae)
TYPE OF TOXIN:
diterpene esters of the tigliane and daphnane types

SYMPTOMS OF POISONING IN HUMANS:
DIGESTIVE: severe oral pain, inflammation, blistering, abdominal cramp and pain, vomiting, bloody diarrhoea
SKIN: burning, reddening, inflammation, blisters
EYES: pain, conjunctivitis, photophobia, temporary blindness

The manchineel tree is found from the coast of Mexico south through Central America to Colombia and Venezuela, as well as in the West Indies and Florida. Its caustic properties soon became known to European explorers of the New World, who encountered the tree on beaches. In the sixteenth century, Oviedo noted its danger in his book on the natural history of the West Indies (which incidentally also included the first illustration of a pineapple):

BELOW **Leafy branches and immature fruit of the manchineel (*Hippomane mancinella*), also known as *manzanita de la muerte*, literally 'little apple of death', are a risk for unwary visitors to tropical beaches.**

It has been proved many times that if men carelessly lie down to sleep under the trees, when they rise after a short nap there is a great pain in the head and swelling of the eyes and cheeks. And if by chance the dew from the tree falls on the face, it is like fire, blistering and burning the skin wherever it touches; and if it falls in the eye it blinds or burns them, and the sight is endangered. If the wood is burned no one can endure it long, for it causes much heaviness, and such headaches that all stand away from it, be they man or any other animal.

Since then, numerous graphic accounts of the symptoms that result from skin or eye contact with the latex of the manchineel tree have been published, so we can be left in no doubt of the harm the species can cause. In addition, it produces deceptively apple-like fruit, which are 3–5 cm (1–2 in) in diameter, and when ripe are yellowish green with flushed red cheeks and an aromatic, pleasant-tasting yellow flesh. These, too, can cause contact reactions, and eating the fruit is even more disastrous, as doing so irritates the mouth, throat and digestive tract; deaths have occurred.

Highly irritant oils

There are several members of the spurge family that are notoriously poisonous, but their toxicity is generally associated with the oil that is obtained from their seeds, which may be caustic, co-carcinogenic or cytotoxic due to the presence of phorbol esters. Such species include the purging nut and the purging croton (*Croton tiglium*); seeds of the latter are the

LEFT **As its leaves emerge in spring, the mezereon or spurge laurel (*Daphne mezereum*) bears fragrant pink flowers directly on its bare stems, which are followed by attractive red fruit.**

intermittent epidemics of livestock disease and death. The syndrome includes chronic diarrhoea and loss of appetite, and affects cattle and sheep. It was described before the Second World War, but the connection to annual species of *Pimelea* was not recognized until the 1970s, and there have been at least two more epidemics in the last 20 years. Studying *Pimelea* was hampered by difficulties in establishing plants in cultivation, as germination of the seeds under laboratory conditions has a very low success rate.

Mezereon (*Daphne mezereum*) is a widespread species of Europe and western Asia, noted for flowering early in the spring before the leaves appear. The scented purple flowers make this shrub a popular ornamental plant despite the fact that it is very poisonous. The fruit are single-seeded, bright red berries that attract thrushes and other birds, which are unaffected by the phorbol esters mezerein and daphnetoxin they contain. Although the berries also entice children, the burning sensation the mezerein causes upon contact with the mouth usually prevents serious poisonings.

BELOW **Pimelea poppy or blood pimelea (*Pimelea haemostachya*), a native of Queensland, Australia, has a reputation for being toxic to livestock, although most cases of *Pimelea* poisoning are associated with other species containing higher concentrations of simplexin.**

source of croton oil. Oil from both of these species was used as a purgative and for other medicinal purposes, and has been investigated for possible anti-HIV (human immunodeficiency virus) and anti-leukaemia properties, among other potential uses. Perhaps the most widespread use of croton oil today is in chemical peels, an alternative to facelift plastic surgery. The croton oil, which is diluted in phenol and water before application, penetrates deep into the skin. The peel has to be performed under sedation in an operating room as it is reported to be very painful, and is associated with a risk of cardiac arrhythmias as well as kidney and liver toxicity.

Parallel evolution

In an evolutionary twist, phorbol esters are not only present in the spurge family, but also in many genera in the mezereum family. This cosmopolitan family contains somewhere between 45 and 50 genera and more than 900 species, most of them woody plants in the form of shrubs or trees. Many have been used in traditional medicine systems or as fish poisons. This is an interesting example of parallelism not only in plant chemistry evolution but also in anthropological use, and attests to the power of phorbol ester activity.

There are a few herbaceous species in the mezereum family. One of the largest genera containing some of these is *Pimelea*, with more than 100 species from Australia and New Zealand. In many of these species a daphnane ester called simplexin is present, which seems to be the cause of

The mustard bomb – isothiocyanates

In plants of the cabbage family and their relatives, an intricate chemical defence has evolved that lies in wait and is activated only upon grazing. It consists of a group of apparently innocuous compounds called glucosinolates and a special enzyme called myrosinase, which are usually separated within the plant cells. They can mix, however, when plant cells are damaged, at which point the glucosinolates are transformed into profoundly irritating isothiocyanate compounds.

About 120 different glucosinolates have been described from almost 4,800 species of plant. More than 95 per cent belong to a group of 17 families, the Brassicales, consisting of the relatives of cabbages (*Brassica oleracea*, with cultivated forms including broccoli, cauliflower, Brussels sprouts and kale; cabbage family), capers (*Capparis spinosa*; caper family, Capparaceae), papaya (*Carica papaya*; papaya family, Caricaceae) and nasturtium (*Tropaeolum majus*; nasturtium family, Tropaeolaceae), while the rest are confined to the putranjiva family (Putranjivaceae).

Some like it hot

When it comes to insects, there are some that have turned glucosinolates and isothiocyanates into an advantage. Every farmer growing cabbage crops fears the large white butterfly (*Pieris brassicae*) and its relatives, whose caterpillars (larvae) feed only on glucosinolate-containing host plants. They have developed a detoxifying mechanism and can ravage cabbage, Brussels sprouts and similar crops, which other butterfly larvae will leave alone. In certain aphids, this adaptation to glucosinolate-containing plants has been taken even further. As they are sap-sucking insects, these aphids seldom activate the mustard bomb of the plant, but instead have invented one of their own! They are preyed upon by hoverflies, and by accumulating glucosinolates from their host plants and producing a myrosinase enzyme of their own, the aphids will produce isothiocyanates when chewed by the hoverfly. The volatile isothiocyanate affects the predators, but also signals to other aphids that a predator is present and it is therefore time to find another feeding ground.

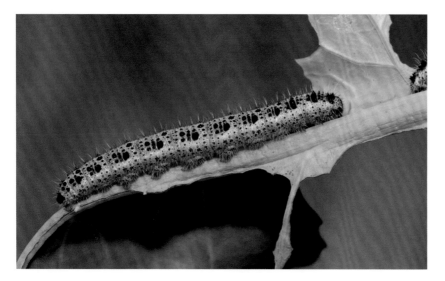

LEFT Caterpillars of the large white butterfly (*Pieris brassicae*) feed on cabbage and its relatives, having evolved a mechanism to detoxify the plants' glucosinolates.

Scary signals

PLANT:
Brassica nigra (L.) W.D.J.Koch, *B. juncea* (L.) Czern., *Sinapis alba* L. and others
COMMON NAMES:
B. nigra – black mustard;
B. juncea – Indian mustard;
S. alba – white mustard
FAMILY:
cabbage (Brassicaceae)
TYPE OF TOXIN:
isothiocyanates

SYMPTOMS OF POISONING IN HUMANS:
NEUROLOGICAL: sweating
DIGESTIVE: sensation of burning and heat in mouth and throat, gastrointestinal irritation, diarrhoea
SKIN: sensation of heat, reddening, itching, swelling, burn-like lesions
EYES: tear production
NOSE: irritation, rhinorrhoea (runny nose)

ABOVE **Flowering Indian or brown mustard (*Brassica juncea*) growing in Thailand, where different forms are used for seed oil or the leaves, stems and roots are eaten as a vegetable.**

Allyl isothiocyanate

LEFT **The volatile allyl isothiocyanate is one of the irritating compounds produced from glucosinolates in mustard plants.**

Several plants in the cabbage family are known by the common name mustard, but the three that are most commonly used as a condiment are the black mustard (*Brassica nigra*), Indian mustard (*B. juncea*) and white mustard (*Sinapis alba*). The leaves of these herbaceous plants, particularly Indian mustard, are used as a vegetable, but it is the small round seeds that are used to make mustard condiment, for which they are usually ground to a powder, or as the source of a culinary oil (see page 175).

The green parts of these plants and their relatives contain fewer irritating compounds than the seeds, but the glucosinolates themselves have a bitter taste that is often off-putting enough to make large herbivores seek other pastures. They are also the reason why some people do not like cooked cabbage or Brussels sprouts, in which the myrosinase enzyme has been denatured by heat, leaving the bitter glucosinolates intact.

When large herbivores graze on these plants, however, they usually destroy the plant matter in a way that sets off the 'mustard bomb'. If you have ever used slightly too much wasabi on your sushi, or fresh horseradish on your Sunday roast, you will be well aware of the deterrent effect of the pungent isothiocyanates. These irritate the mucous membranes in the mouth, nose and sinuses due to their activation of receptors related to those affected by capsaicin (see pages 122–123).

Life-saving irritation

Humans have traditionally used the irritating isothiocyanates for a range of medicinal applications. One of these is as a 'mustard plaster', which is used to treat rheumatism, aching joints, painful muscles and chest infections. Ground seeds of one of the mustard species are mixed with flour into a paste that is then spread on a soft cloth and applied to an area of the body. The mustard paste should not touch the skin, and the plaster should be applied for only a short time – 15 minutes or so. Severe burns to the skin have resulted from direct application of the mustard powder and prolonged exposures.

The humorous 2011 Ig Nobel Prize for Chemistry went to research into a novel use for the highly irritating action of the wasabi volatile isothiocyanate. The wasabi 'odour-generation alarm' would wake a sleeping person in the event of a fire by irritating the mucous membrane of their nose.

RIGHT **White mustard (*Sinapis alba*), showing the cruciform flowers with four petals that is a feature of the cabbage family.**

Heat without temperature – capsaicin

The fruit and seeds of species of chilli peppers (*Capsicum* spp.) in the potato family (Solanaceae) contain a pungent compound, capsaicin, that makes food 'hot'. As well as being used by humans as a spice for thousands of years, capsaicin also has medicinal applications, and the burning discomfort and pain it causes have found roles in riot-control and self-defence.

Illusions of heat

PLANT:
Capsicum annuum L.,
C. frutescens L. and
C. chinense Jacq.

COMMON NAMES:
C. annuum – chilli pepper, cayenne pepper, paprika, bird pepper, red pepper, green pepper, bell pepper;
C. frutescens – tabasco pepper, chilli pepper, cayenne pepper; *C. chinense* – bonnet pepper, habanero pepper

FAMILY:
potato (Solanaceae)

TYPE OF TOXIN:
capsaicin and other capsaicinoid alkaloids

SYMPTOMS OF POISONING IN HUMANS:
NEUROLOGICAL: sweating
DIGESTIVE: sensation of burning and heat in mouth and throat, gastrointestinal irritation, diarrhoea
SKIN: sensation of heat, reddening
EYES: tear production, burning discomfort, pain

The chilli or chili pepper (*Capsicum annuum*) is a small shrub from Mexico and Guatemala, with simple leaves and pendant, star-shaped flowers that appear singly and are followed by elongated, brightly coloured fruit. Numerous cultivars have been bred that vary in the size, shape and pungency of these fruit. They include the large, sweet bell peppers, as well as mild to hot chilli peppers. The taxonomy of chillies is complicated, however, with some cultivars of *C. annuum* having characteristics that overlap with those of two other species, the Tabasco pepper (*C. frutescens*) from Bolivia and western Brazil, and the very hot bonnet pepper (*C. chinense*), which despite its specific epithet is from Bolivia, northern Brazil and Peru. Some prefer to treat these three species and their cultivars as the '*annuum–chinense–frutescens* complex'.

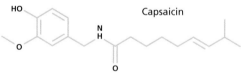

ABOVE **Capsaicin initiates a burning sensation, but will then inhibit nerve signalling and cause numbness instead.**

LEFT **To alleviate the 'heat sensation' from chilli, try eating a yogurt raita containing chopped mint (*Mentha* spp.) leaves, as the menthol from the mint stimulates 'cold sensation' neurons.**

ABOVE Chilli pepper plants (*Capsicum annuum*), with fruit ripening from green to red. Plants thrive in warm, moist conditions and form branching shrubs up to 1.5 m (5 ft) tall.

Scoville's test

The pungency of chilli peppers varies between the numerous cultivars that have been bred, and it was in 1912 that Wilbur Scoville (1865–1942), an American pharmacist, devised a method to measure this variation. The Scoville organoleptic test, which measures this pungency in Scoville heat units (SHUs), relies on a panel of expert tasters to determine how dilute a solution of chilli extract in sugar water has to be before it cannot be detected. Pungency varies from 0 SHU for bell peppers, to more than 1.5 million SHU for the hottest chilli, the cultivar *Capsicum chinense* 'Carolina Reaper', which entered the *Guinness Book of Records* in 2013. To put this in perspective, jalapeño peppers have a score of 2,500–8,000 SHU. Although the test was devised for chilli peppers, it can also be applied to other pungent compounds. And while pure capsaicin has a score of 16 million SHU, the winner of the Scoville test is actually resiniferatoxin, a diterpene from the resin spurge, with a score of 16 billion SHU (see box, page 116).

ABOVE The 'Carolina Reaper', or HP22B Chilli Pepper, was bred in the United States. With a pungency of 1.5 million SHU, it was recognized as the world's hottest pepper in 2013.

SPICING THINGS UP

Some culinary traditions use more chilli pepper than others, with the highest number being eaten in the species' native Mexico (one chilli per person per day). Chilli has also been embraced in many of the countries to which it has been introduced, particularly India, where it is a key ingredient in curries, and Thailand. Either the fresh fruit and seeds, or the powdered or flaked dried fruit, are used for seasoning during cooking or as a condiment.

The pungent compounds in chilli peppers, including capsaicin (8-methyl-*N*-vanilloyl-6-nonenamide), are capsaicinoid alkaloids, which bind to vanilloid receptors on sensory neurons (known as transient receptor potential vanilloid (TRPV) channels). These same receptors can also be stimulated by heat and pain, so the binding of the capsaicin results in the sensation of heat. The degree of burning and reddening is related to the concentration of capsaicinoids (see box) and duration of exposure (a dose-related response). TRPV channels are common to all mammals, and thereby deter rodents and other mammalian pests from eating chilli crops. Birds lack the capsaicin-binding site of these channels, however, so eat the ripe red fruit and disperse the seeds without harm.

In addition to the sensation of heat and burning in the mouth, eating large amounts of hot chillies can cause irritation of the gastrointestinal tract. It is the burning discomfort and pain that chillies or concentrated chilli extracts cause to the eyes and nose that can be most distressing. Pepper sprays have proved to be effective weapons since they were first employed by Mayan Indians, and police forces in a number of countries now use them in the control of unruly individuals and crowds. However, the legality of using pepper sprays for self-defence varies around the world.

RANUNCULACEAE

The buttercup family is distributed globally but is found mainly in temperate and northern regions, where there can be few gardens that do not include a member, even if it is just a buttercup in the lawn. It is considered to be among the oldest families and consists of more than 2,500 species in 62 genera. Most species are either annual or perennial herbs or climbers, or occasionally shrubs. They vary in form, especially in the appearance of their flowers, which can be simple, such as in buttercups (*Ranunculus* spp.) or clematis (*Clematis* spp.), or complex, for example in granny's bonnets (*Aquilegia* spp.) or aconites (*Aconitum* spp.; see pages 48–49).

The main toxic compounds in members of the buttercup family are the aconitine-type diterpene alkaloids, cardiac glycosides and ranunculin. The diterpene alkaloids are found principally in species of aconite, and also larkspurs (*Delphinium* spp.) (see pages 48–49). They stimulate and then depress the central nervous system and peripheral nervous system, finally causing death by heart failure. Both types of cardiac glycoside occur in the buttercup family, with cardenolides being found in species of pheasant's eye (*Adonis* spp.; see page 58) and bufadienolides in some species of hellebore (*Helleborus* spp.), including the Christmas rose or black hellebore (*H. niger*; see page 60). The Christmas rose is also one of the species of hellebore to contain ranunculin, the precursor to the irritant protoanemonin, which is also found in several other genera in the family, including the anemones (*Anemone* spp.) (see pages 126–127).

BELOW RIGHT **The creeping buttercup (*Ranunculus repens*) has the classic buttercup flower, with five simple yellow petals. Grazing animals avoid buttercups due to the presence of ranunculin.**

BELOW **Flowers of the common columbine (*Aquilegia vulgaris*), native to Europe, have five spreading, petal-like sepals and five petals with backward-pointing nectar-filled spurs.**

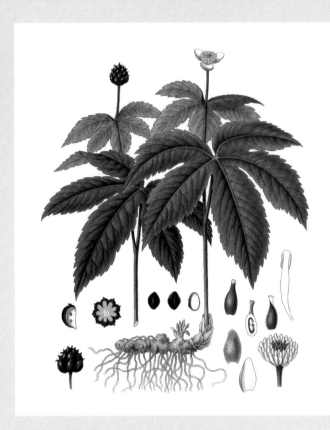

ABOVE **Goldenseal (*Hydrastis canadensis*)** was used by Native Americans to treat sore eyes; its current popularity as a medicinal plant has seen wild populations decline.

MEDICINALLY RELEVANT

The plants containing these toxic compounds are used in various traditional systems of medicine, although in the Western world this has decreased due to concerns about their potential for harm. There are also many other species in the buttercup family that have medicinal applications but are not poisonous enough to be described in detail in this book. These include goldenseal (*Hydrastis canadensis*), Chinese goldthread (*Coptis chinensis*) and meadowrue (*Thalictrum* spp.), which, among other medicinally active compounds, also produce alkaloids that are found in other families, particularly the poppy (Papaveraceae) and barberry (Berberidaceae) families. Remedies that combine goldenseal with echinacea (*Echinacea* spp., in the daisy family (Asteraceae); see page 160) to boost the immune system are one of the top-selling botanical dietary supplements in the United States. And black cohosh (*Actaea racemosa*, syn. *Cimicifuga racemosa*) is used to treat symptoms of menopause, although it has been implicated in a few cases of liver disease; Asian species of *Actaea* are used for other conditions in traditional Chinese medicine.

FOOD FOR OUR ANCESTORS

Very few members of the buttercup family are used as food – nigella or black cumin (*Nigella sativa*), whose seeds are used as a spice, is one of the only species grown on a commercial scale. Lesser celandine (*Ficaria verna*, syn. *Ranunculus ficaria*) is a small perennial that is native to most of Europe, including Scandinavia. It is one of the first plants to come into leaf in spring, bearing bright yellow flowers before dying back in the summer and overwintering as small tubers close to the soil surface. The young leaves are used today, raw or boiled, in traditional vegetable mixtures in Italy, as is traveller's joy (*Clematis vitalba*), which is boiled before eating.

There is also evidence that our ancestors ate the tubers of lesser celandine. Archaeological excavations in northern Germany, Denmark and elsewhere in Europe have uncovered charred lesser celandine tubers dating to the Mesolithic (9000–4000 BC) and Neolithic (4000–1800 BC), particularly the Early Neolithic (4000–3400 BC). They were found in domestic contexts, indicating that the tubers were used as a food, rather than the charring occurring as a result of vegetation burning. Cooking or roasting and grinding the tubers would destroy any protoanemonin that might be produced from the ranunculin in the tubers, making them safe to eat. They were a useful source of starch at a time when humans were more reliant on food collected in the wild.

RIGHT **Lesser celandine (*Ficaria verna*), with 8–12 petals**, should not to be confused with greater celandine (*Chelidonium majus*; see page 173) in the poppy family.

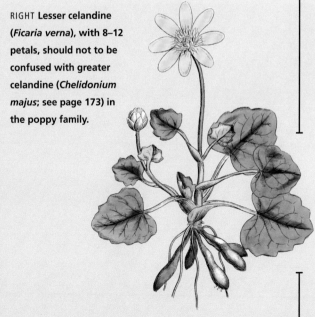

An immediate irritant – protoanemonin

Many members of the buttercup family (see pages 124–125) can cause chemical irritant contact dermatitis when fresh due to the presence of the protoxin ranunculin, which is restricted to this family. This irritancy is said to have been used by beggars in ancient Rome to produce skin blisters and hence make themselves more pitiable. In addition to buttercups themselves, other family members that produce ranunculin include species of anemone and pasqueflower (*Anemone* spp., syn. *Pulsatilla* spp.), marsh marigold (*Caltha* spp.), bur buttercup (*Ceratocephala* spp.), clematis, lesser celandine and hellebore.

Activation of the toxin

Ranunculin, a terpenoid glucoside, is stored in the cell vacuole, but if the plant is damaged, the cell contents mix together and the ranunculin comes into contact with the enzyme *beta*-glucosidase. This splits the sugar (glucose) from the ranunculin, creating protoanemonin, a volatile unsaturated lactone of *gamma*-hydroxyvinylacrylic acid (4-hydroxypenta-2,4-dienoic acid), which is highly reactive. As the plant material dries, protoanemonin converts (dimerizes with itself) to the non-reactive anemonin; it is also destroyed by high temperatures.

BELOW **Collecting Christmas rose (*Helleborus niger*) seeds when the fruit capsules are still green can irritate the fingers of horticulturists and gardeners due to the presence of ranunculin.**

ABOVE **The bur or curveseed buttercup (*Ceratocephala testiculata*) reaches only 8 cm (3 in) in height. Its heads of 5–80 curved seeds form a bur that hooks onto fur and clothing, facilitating their dispersal.**

The volatile protoanemonin gives freshly crushed plant material a pungent smell that can irritate the eyes and nose. Handling or eating such plant material irritates the skin, or mucous membranes and digestive tract, respectively. On contact, protoanemonin can combine with sulfhydryl (-SH) groups in the skin, and disrupt the disulfide bonds in, for example, proteins and glutathione, therefore causing cell death (cytotoxicity). It can also alkylate DNA and cause genetic mutations. The resulting effects depend on the duration and extent of contact, and the amount of protoanemonin released by the particular plant. They can range from warmth, burning and irritation, to swelling, blistering and ulceration.

A POULTICE TO AVOID

PLANT:
Ceratocephala testiculata (Crantz) Besser (syn. *Ranunculus testiculatus* Crantz) and *C. falcata* (L.) Pers. (syn. *R. falcatus* L.)

COMMON NAMES:
C. testiculata – bur buttercup, curveseed buttercup;
C. falcata – beakwort

FAMILY:
buttercup (Ranunculaceae)

TYPE OF TOXIN:
unsaturated lactone (protoanemonin)

SYMPTOMS OF POISONING IN HUMANS:
DIGESTIVE: irritation of the mouth and throat, nausea, vomiting, bloody diarrhoea, liver toxicity
SKIN: painful reddening, blistering, ulceration

Ranunculin

Protoanemonin

ABOVE **The glycoside ranunculin transforms into the reactive protoanemonin if the plant tissue is damaged. It causes irritation in a herbivore's mouth, deterring further feeding.**

The bur buttercups (*Ceratocephala* spp.) are a small genus in the buttercup family, found in southern Europe, western Asia and north Africa. They are used traditionally as a treatment for rheumatism and wound healing in Turkey. However, on a significant number of occasions use of the plants has resulted in serious effects to the treated area, such as extensive blistering, chemical burns and open wounds. Such cases involve the use of poultices made from finely chopped and ground bur buttercup to treat joint pain, applied under a closed dressing for between two hours and two weeks. This method both maximizes the creation of protoanemonin from ranunculin, and delays its conversion to the less irritant anemonin.

The immediate irritancy of protoanemonin is a very effective anti-feedant, so both insects and larger animals (including humans) are deterred from eating plants that produce this toxin. It is noticeable, for example, that buttercup plants in fields are usually left untouched by grazing animals. On occasion, however, sheep have been killed by eating bur buttercups, and cows by eating buttercups. Animals are most at risk when there is little else to eat.

Doctrine of signatures

The concept that organisms were created to fulfil specific purposes was introduced in Chapter 1 (see pages 16–17). Going beyond that, the doctrine of signatures suggests that nature, or a divine being, grants a 'signature' to those plants that have a particular medicinal property. This can be recognized by a similarity between the form of the plant, such as its shape, colour, taste or odour, and the part of the human body or the condition that it treats. Some also draw parallels between the 'essence' of the plant and the personality of the patient or psychological dimension of the disease. Versions of the doctrine have existed from early times in many medicinal traditions, including Indian and Chinese systems. Within Europe, the doctrine of signatures particularly flourished in the Middle Ages, when physicians were actively studying medicinal plants and human ailments. The Swiss alchemist Paracelsus (1493–1541) is credited with raising the theory's status from that of folklore to a seriously considered doctrine.

Examples of such signatures can be found in the buttercup family. The tubers of the lesser celandine (see box, page 125) resemble piles, or haemorrhoids, and were therefore used to treat that condition, giving the species the alternative common name of 'pilewort'. The leaf shape of the liverwort (*Anemone hepatica*, syn. *Hepatica nobilis*) resembles the liver and was used to treat that organ. Nowadays, the doctrine of signatures persists to some extent in complementary and alternative medicine systems such as homeopathy and naturopathy.

BELOW **Liverwort (*Anemone hepatica*), from woodlands in Europe and North America.**

Blisters and burns – furanocoumarins

The light-induced plant dermatitis known as phytophotodermatitis is caused by a number of plants, including many members of the carrot family (Apiaceae; see pages 74–75). Contact with these plants does not always result in dermatitis, however, as conditions need to be right. Levels of the phototoxic furanocoumarins vary between plant parts, being greatest in the developing seeds, and as their primary role is probably defensive, they also increase in response to environmental stress such as drought or infection by pathogenic fungi. In addition to contact with phototoxins, exposure to sufficient ultraviolet (UV) radiation is required for dermatitis to occur, which is particularly limiting in temperate regions.

LEFT **Giant hogweed (*Heracleum mantegazzianum*) is an invasive weed that competes with native plants in parts of Europe and North America. Many countries now take steps to control its spread.**

Horrible hogweeds

PLANT:
Heracleum mantegazzianum Sommier & Levier and other *Heracleum* spp.
COMMON NAMES:
giant hogweed
FAMILY:
carrot (Apiaceae)

TYPE OF TOXIN:
furanocoumarins (psoralens)
SYMPTOMS OF POISONING IN HUMANS:
SKIN: burning, reddening, small to large blisters, burn-like lesions, hyperpigmentation

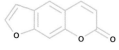
Psoralen

LEFT **Under the influence of ultraviolet light, the furanocoumarin psoralen becomes reactive, causing severe skin blistering.**

The giant hogweed (*Heracleum mantegazzianum*) is a spectacular plant from southern Russia and Georgia. Growing to a height of 3.5 m (11 ft) and bearing thick purple-spotted stems, enormous toothed leaves and flower umbels measuring up to 60 cm (2 ft) across, this striking plant was commonly grown as an ornamental in the nineteenth century before its sinister side became more widely known. Together with closely related species such as Sosnowsky's hogweed (*H. sosnowskyi*), from Georgia, Azerbaijan and Armenia, giant hogweed is still grown as a forage crop for livestock. Such human intervention means

Berloque dermatitis and Margarita photodermatitis

Phytophotodermatitis has had a number of different names since it was first observed, prompted by some unusual routes of exposure to the psoralens in citrus peel. Berloque dermatitis results from repeated exposure to weak dilutions of funanocoumarins in perfume and similar products. The main source of these is bergamot oil from the bergamot orange (*Citrus bergamia*), which contains bergapten (5-methoxypsoralen, 5-MOP). Bartenders are at greatest risk of developing Margarita photodermatitis, also known as lime disease (not to be confused with Lyme disease), which can occur after hand-squeezing lime or lemon juice in bright sunshine, such as when preparing fruit cocktails in tropical countries.

LEFT Bergamot orange (*Citrus bergamia*) is the source of bergamot oil, used to flavour Earl Grey tea.

---

that it is now found across much of Europe, usually in damp areas such as riverbanks, drainage ditches and road verges.

A lack of awareness by the public of the dangers of touching giant hogweed results in sporadic cases of phytophotodermatitis, or small outbreaks in particularly sunny weather. Country walks and children playing outdoors – normally healthy pursuits – can result in serious reactions. Symptoms are delayed, with a burning sensation starting about 24 hours after exposure to the sun, and blistering a further 24 hours later. Classic routes of exposure to other psoralen-producing members of the carrot family include clearing weeds such as bolted parsnips (*Pastinaca sativa*) from vegetable plots and weeding between rows of celery (*Apium graveolens*) with bare hands, or strimming (string trimming) verges of cow parsley (*Anthriscus sylvestris*) while stripped to the waist.

Members of the citrus family can also cause phytophotodermatitis, including citrus (*Citrus* spp.) fruits and their essential oils (see box), and common rue (*Ruta graveolens*; see page 109). Not only have plants of common rue caused serious blistering to unsuspecting gardeners and children, but reactions have also occurred when the crushed leaves or infusions have been rubbed on skin with the intention of repelling insects or relieving pain.

EXTERNAL EFFECTS OF INTERNAL TREATMENTS

Skin conditions such as psoriasis and vitiligo have long been treated using psoralens from plants, such as the seeds of bakuchi (*Cullen corylifolium*, syn. *Psoralea corylifolia*; pea family) and bishop's weed (*Ammi majus*; carrot family), with the earliest known records dating from around 1400 BC in India. Originally applied externally but now also available in a form that is taken internally, these treatments are termed PUVA (psoralen followed by exposure to ultraviolet A, or UVA). When used correctly, the psoralen increases the sensitivity of the skin to UVA, which in turn – in the case of psoriasis – reduces the inflammation in the skin. However, exposure to the wrong wavelengths of light, or for too long, has resulted in severe, and occasionally life-threatening, sunburn. Eating large quantities of a psoralen-containing plant followed by exposure to UV light can also cause phototoxic effects.

Other phototoxins – such as fagopyrin in buckwheat (*Fagopyrum esculentum*; dock family, Polygonaceae) and hypericin in St John's wort (*Hypericum perforatum*; St John's wort family, Hypericaceae) – can also cause skin reactions if eaten in large quantities. Once known only in animals, such as sheep grazing on species of *Hypericum* in Australia, cases are now occasionally seen in people who juice buckwheat shoots. Taking St John's wort supplements is unlikely to result in symptoms unless normal doses are greatly exceeded.

BELOW Seeds of bishop's weed (*Ammi majus*), an annual plant from the Mediterranean region, including north Africa, are a rich source of psoralen, used to treat vitiligo.

Poison ivy

Of the 80 genera in the largely tropical cashew family (Anacardiaceae) of trees, shrubs and lianas, a third are known to include species that produce allergenic oleoresins, of which there are two types: alkylcatechols and alkylresorcinols. The most allergenic of the compounds, urushiol, is a mixture of alkylcatechols and is found in a number of plants from several genera, including poison ivy (*Toxicodendron* spp.). Alkylresorcinols are found in commercially important plants such as cashew (*Anacardium occidentale*) and mango (*Mangifera indica*). The allergenic oleoresins fulfil an anti-feedant role in the plants that produce them. For example, plants in the cashew family are generally a food source for hemipteran insects, but these insects do not feed on species that produce alkylcatechols. In humans, the allergens cause a type IV (delayed) hypersensitivity reaction.

A GUARANTEED ALLERGY

PLANT:
Toxicodendron pubescens Mill. (syn. *Rhus toxicodendron* L.), *T. radicans* (L.) Kuntz (syn. *Rhus radicans* L.) and others

COMMON NAMES:
T. pubescens – poison ivy, poison oak; *T. radicans* – poison ivy, eastern poison ivy

FAMILY:
cashew (Anacardiaceae)

TYPE OF TOXIN:
alkylcatechols (urushiol)

SYMPTOMS OF POISONING IN HUMANS:

SKIN: papulovesicular rash (with both solid and fluid-filled patches), often in linear streaks; swelling, weeping, crusting, severe itching

OTHER: occasionally fever, tiredness, weakness

Species of poison ivy and the related poison oaks and sumacs (*Toxicodendron* spp.) are the most common cause of acute allergic contact dermatitis in the United States. They are found in every state except Hawai'i and Alaska, with different species being more abundant in different regions. In addition, the genus occurs in South America and temperate and tropical Asia. In most other countries, you are unlikely to come into contact with one of these plants, except perhaps in a botanic garden. The closely related *Rhus* genus, to which species of *Toxicodendron* have at times been referred, includes ornamental plants such as stag's horn sumac (*R. typhina*). Urushiol is absent from *Rhus* species but they do produce far less allergenic biflavonoids.

Poison ivy and its relatives are climbers, shrubs and small trees. They have compound leaves with three or more leaflets, and bear insignificant flowers and pale fruit. Urushiol is found in all parts of the plant and in the sap, which is a colourless or slightly yellow, oily resin that turns black when exposed to the air. Although not volatile (it remains a liquid even at 315 °C, or 600 °F), the oil attaches to smoke particles if burnt, such as in a forest fire or on a campfire. It can also attach to the fur of animals, including pets. These are therefore routes by

ABOVE **Urushiol is actually a mixture of several similar compounds, all with the potential to cause painful allergic reactions after repeated exposure.**

LEFT **'Leaves of three, let it be' is a rhyme taught in the United States to help people distinguish poison ivy (*Toxicodendron radicans*) from similar-looking plants.**

ABOVE **Poison ivy (*Toxicodendron radicans*) leaves have attractive autumn colour and the plants are occasionally grown as ornamentals for that reason, putting gardeners in particular at risk.**

which the allergens can be transferred to humans without them even having to touch the plants themselves – the usual route of exposure.

The first time a person is exposed to urushiol, symptoms can take 10–14 days to develop and are relatively mild. However, once a person is sensitized, subsequent exposures give rise to symptoms within 24–48 hours. The sap is absorbed quickly by the skin and can also be inadvertently transferred from the point of contact with the plant to other areas of the body. The resulting rash is extremely itchy and almost impossible not to scratch. Symptoms usually resolve within two to three weeks, unless there is a second exposure in the meantime. The percentage of Americans sensitive to the extremely allergenic urushiol has been estimated at up to 80 per cent. An unlucky 10–15 per cent are exquisitely sensitive, developing a severely itchy and extensive rash within two to six hours, accompanied by fever, tiredness and weakness.

Culinary hazards

Native to northern South America, the cashew tree is now widely cultivated in tropical areas. Its small kidney-shaped fruits each contain a single seed, the cashew nut, and hang below the cashew 'apple', which swells as the seed ripens. The shell around the nut causes allergic dermatitis, being particularly high in allergenic compounds. The process of extracting the cashew nut from the shell is hazardous for workers, who develop chronic dermatitis on unprotected hands, and who are also exposed to the fumes from roasting nuts, resulting in ulceration and scarring in the mouth. Eating insufficiently processed nuts can also result in dermatitis that affects the skin around the mouth and elsewhere on the body. The risk is particularly high for those who are already sensitized to the urushiol in poison ivy, as there is cross-sensitivity to the resorcinols in cashew. Such cross-sensitivity also occurs with the resorcinols in the peel of the mango fruit, and the pink peppercorns of the Brazilian pepper tree (*Schinus terebinthifolius*).

BELOW **The fleshy cashew (*Anacardium occidentale*) 'apple' is actually the swollen fruit stem, while the single-seeded fruit, with its distinctive cashew-nut shape, hangs below.**

CHAPTER 7

GUTS WITHOUT THE GLORY

Many plants cause discomfort as a means of deterring herbivores from continuing to feed, and for most people nausea, vomiting and diarrhoea are quite effective at preventing further consumption of these plants. This chapter shows that such effects might not only be a way to expel toxic substances, but may also be an early warning sign of potentially lethal poisonings.

GUT REACTIONS: MECHANISMS OF ACTION

There are some symptoms that are more or less common in all cases of poisonings, whether accidental or intentional. There is usually a quickening of the heartbeat and breathing, a shaky feeling, as well as nausea and stomach ache. As these symptoms may simply reflect the mental state of a person – we all display one or all of them right after we have had a scare, for example – they are too general to use for diagnosing a poisoning. However, although vomiting and diarrhoea are usually harmless, in this chapter we will meet the plants that have taken these effects to the next level, causing extended or intensive gastrointestinal distress, with potentially lethal fluid and salt loss.

NAUSEA AND VOMITING

Feeling sick is probably something we have all experienced, and sometimes it is followed by vomiting. The mechanism of vomiting is understood fairly well and involves two connected centres in the brainstem: the chemoreceptor trigger zone (CTZ) and the vomiting centre. The CTZ cannot induce vomiting by itself, but influences the vomiting centre by stimulation, which causes vomiting, or by inhibition, which prevents it. The CTZ is activated by changes in the blood, such as high levels of urea due to kidney failure, but will also respond to antiemetic drugs.

The vomiting centre is not only regulated by the CTZ, but also receives nerve signals from several organs such as the stomach and intestines, and sensory input, for example in the form of balance, smell and psychological stimuli such as fear. When stimulated, it initiates several actions: a reversal of peristalsis in the small intestine, contraction of the abdominal muscles and parts of the stomach, and finally evacuation of the stomach contents through the mouth.

Plants that cause vomiting may do so by activating the CTZ, such as ipecacuanha (see pages 136–137); by physically irritating the stomach lining, such as the glycoalkaloids in potatoes and their relatives (see pages 140–141); or by a combination of the two, as with alkaloid-containing species in the amaryllis family (Amaryllidaceae) (see pages 142–143).

ENTERIC EXPULSIONS

Once ingested, material passes through the stomach and enters the intestines. The body's natural function is to absorb the nutrients in food before excreting the waste as faecal matter. In the small intestines, where most of the uptake of compounds takes place, a large volume of water is added to the mix, which is then reabsorbed in the large intestine. If this reabsorption process is not working, the sheer volume of the intestinal contents will cause diarrhoea. This may occur when the compounds that are not absorbed prevent reabsorption of the water by osmosis, as is the case for certain carbohydrates and proteins, or when there is an increase in the secretion of water into the intestine.

Diarrhoea can also result from an increase in bowel movement (peristalsis), where the flow of faecal matter is greater than the absorption rate of its water content. This can be due to a number of factors stimulating the activity of the enteric nervous system. The most well known of these are a change in the composition of the intestinal microorganisms, direct stimulation of nerve signalling and irritation or even death of the cells in the mucosal lining of the large intestine.

It is often difficult to assign a single mechanism to diarrhoea, and sometimes the root causes are not well studied.

LEFT **Glycoalkaloids are found in the *Solanum* genus, including the silverleaf nightshade (*S. elaeagnifolium*). This native of the southwestern United States and Mexico is now found in many semi-arid regions of the world.**

ABOVE **The bark of alder buckthorn (*Frangula alnus*, syn. *Rhamnus frangula*), a small tree from Europe, north Africa and western Asia, is used as a stimulant laxative.**

The laxative anthraquinones (see pages 144–145) probably act through stimulation of the nervous system, causing increased peristalsis, but may also affect the reabsorption of water. Ricin and other lectins (see pages 146–149) probably act by causing inflammatory cell death in the intestinal wall, but may also disturb the microorganism composition. The mechanism of action for cucurbitacins is unclear, though their ability to affect cell division will at least be part of their effect (see pages 150–151). The very potent alkaloid colchicine inhibits cell division, preventing the replacement of cells and stopping cell functions (see pages 152–153). This causes devastating cell death in the lining of the intestinal wall due to the very high proportion of actively dividing cells in this tissue, and prevents both reabsorption of water and bowel movement.

Ipecac

Nausea, vomiting and diarrhoea are common symptoms experienced after a very wide range of poisonous plants and other toxic agents are ingested. However, only one plant has become famous for its use in inducing these symptoms on purpose: ipecacuanha (*Carapichea ipecacuanha*). A syrup or tincture known as ipecac, prepared from the roots of this plant, received wide use as a treatment in acute poisonings, though nowadays this is mostly discontinued due to the associated risks and inherent toxicity.

THE SICK-MAKING PLANT

PLANT:
Carapichea ipecacuanha (Brot.) L. Andersson (syn. *Cephaelis ipecacuanha* (Brot.) Willd., *Psychotria ipecacuanha* (Brot.) Stokes, *Uragoga ipecacuanha* (Brot.) Baill., *Cephaelis acuminata* H. Karst.)
COMMON NAMES:
ipecacuanha
FAMILY:
coffee (Rubiaceae)
TYPE OF TOXIN:
tetrahydroisoquinoline monoterpene alkaloids (emetine, cephaeline and others)

SYMPTOMS OF POISONING IN HUMANS:
After a single ingestion:
DIGESTIVE: nausea, diarrhoea, violent vomiting
NEUROLOGICAL: sweating
After repeated doses:
GENERAL: dehydration, electrolyte imbalances
CIRCULATORY: arrhythmias, heart failure
MUSCLES: weakness

LEFT The alkaloid emetine induces vomiting but should not be used as a first aid after poison ingestions.

The ipecacuanha plant is a small shrub with opposite, simple leaves that have fringed stipules, and dense heads of small white flowers that develop into a two-seeded berry-like fruit. The storage roots of the plant have a distinct cork-like bark. Ipecacuanha grows in Central and South America, and derives its name from the Tupi word *i-pe-kaa-guéne*, which means 'roadside sick-making plant'. The active compounds, the alkaloids emetine and cephaeline, have only been isolated from this plant and a few of its close relatives

RIGHT **Ipecacuanha (*Carapichea ipecacuanha*) plant, with pairs of glossy leaves and, at its tip, a head of small white flowers still in bud and surrounded by four large leaf-like bracts. In the Mato Grosso region of Brazil, the *poayeros* (collectors) took care to replant some roots to ensure another crop could be collected a few years later.**

in the genus *Carapichea*. Several other plants that do not in fact contain any of these emetic alkaloids have nevertheless become known as ipecac. Striated ipecac (*Ronabea emetica*), also in the coffee family, was imported to Europe in large quantities until it was shown at the end of the nineteenth century that it did not actually contain any emetine or cephaeline.

Killing amoebic killers

Ipecacuanha was introduced to Europe at the end of the seventeenth century, where it became famous after curing the son of Louis XIV of France of dysentery in 1686. This intestinal disease is loosely defined as 'bloody diarrhoea' and can be caused by a number of infectious agents. One of the most common of these is the parasitic amoeba *Entamoeba histolytica*, and such amoebic dysentery still resulted in an estimated 55,000 deaths worldwide in 2010.

The alkaloid emetine is effective in killing *Entamoeba histolytica* and, even though this reason for its effectiveness was not known in the seventeenth century, it soon became the drug of choice for any diarrhoeal disease (the recognition that microbes such as amoeba and bacteria cause disease was not established until the nineteenth century). The treatment, however, is not without risks of its own, as the water and salt losses from vomiting can add to the effects caused by the disease itself and make the condition worse. It is also an unsuitable treatment due to the extremely long elimination time of the alkaloid – up to one or two months. Consequently, repeated doses can soon lead to toxic effects, including muscle weakness and heart failure, even if only small amounts are used each time.

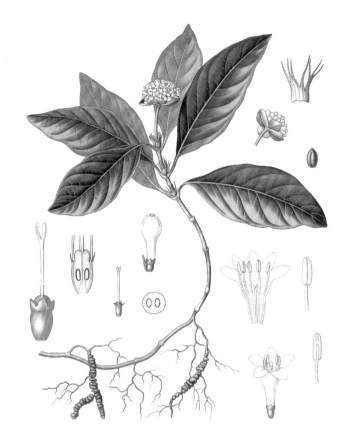

ABOVE **Illustration of the ipecacuanha (*Carapichea ipecacuanha*) plant, showing, among other things, the storage roots with their cork-like bark and the different parts of a flower.**

Poison by proxy

In 1977, the term Munchausen syndrome by proxy was first applied to two cases of child abuse, where the parents had wilfully caused harmful effects to their children, resulting in the children needing hospital care. Although neither of these incidents involved the use of ipecac, overviews of subsequent reported cases of the syndrome show that it has been a common agent. The production of acute vomiting and diarrhoea are non-specific symptoms and ipecac alkaloids are not routinely tested for, so progression to muscle symptoms, including the cardiac muscles (cardiomyopathy), are a real risk. This is another reason why ipecac remedies are no longer recommended as over-the-counter drugs for first aid.

Not a first aid

Ipecacuanha has in modern times been available as a liquid extract and as a syrup, both used to induce vomiting as a first aid after the ingestion of other poisons. It was previously recommended for home use, especially in households with toddlers and young children, but is now considered to be obsolete as the risks associated with ipecac and vomiting are greater than its effectiveness in preventing poisonings. As the liquid extract is more potent than the syrup, unintended overdoses due to mix-ups between the different preparations have sometimes produced severe poisonings, with vomiting so violent that it has caused hernias and stomach ruptures, and even deaths. There are also several described cases of abuse of ipecac by bulimics, in which the combination of malnourishment, vomiting and the development of heart failure and muscle toxicity after extended use led to irreversible and fatal damage.

LEFT *Ipecacuanhae Radix* **is the pharmacopoeia name for the dried roots of ipecacuanha (*Carapichea ipecacuanha*), the part of the plant that is used medicinally.**

SOLANACEAE

This large, almost global family of 2,500 species in more than 100 genera includes many plants that are edible, medicinal and ornamental. However, it also includes species that contain toxic alkaloids, resulting in a wide range of unpleasant and even deadly symptoms. The darker side of this family, exploited by witches and murderers, as well as recreationally by millions, is explored throughout this book.

FRIEND AND FOE

It is hard to imagine a meal that does not contain at least one member of the potato family. Not only does it include one of the world's staple root crops, the potato (*Solanum tuberosum*), but other species have edible fruit that are valued as vegetables – the tomato (*S. lycopersicum*, syn. *Lycopersicon esculentum*), aubergine or eggplant (*S. melongena*), bell and chilli peppers (*Capsicum* spp.; see pages 122–123) – or eaten as fruit, such as cape gooseberry (*Physalis peruviana*) and the 'superfood' goji berry (*Lycium chinense*).

BELOW **Black henbane (*Hyoscyamus niger*) is a pungent plant covered in sticky hairs. The fruit are cup-shaped capsules that open at the top when dry to release the numerous seeds within.**

There is, however, a second, more sinister, common name used for Solanaceae: the nightshades. Several plants within the family are also called nightshade, most of which are in the genus *Solanum* and contain the same toxins, glycoalkaloids (see box and pages 140–141). The name deadly nightshade is sometimes misapplied to species in the genus that have small black berries, such as black nightshade (*S. nigrum*), but the true deadly nightshade (*Atropa bella-donna*) instead contains tropane alkaloids, as do plants from several other genera in the potato family (see below and pages 80–83).

DR CRIPPEN

In addition to deadly nightshade, mandrake (*Mandragora* spp.), thornapple or jimsonweed (*Datura* spp.), angel's trumpet (*Brugmansia* spp.) and henbane (*Hyoscyamus* spp.) are all plants that contain tropane alkaloids. Scopolamine (hyoscine), extracted from henbane, is today used as a remedy for sea sickness and to dry up saliva, but in the past it had a wider range of applications, including pre-medication before surgery, to relieve pain in childbirth and to treat mania.

The most infamous use of scopolamine was by Hawley Harvey Crippen (1862–1910), an American who, despite not being qualified as such, called himself Dr Crippen. He moved to Britain in 1900 with his second wife, Cora, whose extravagant lifestyle meant they were always short of money.

LEFT **Hawley Crippen was the London representative for the American firm Munyon's Homeopathic Remedies. He claimed that he had bought scopolamine for use in the preparation of his homoeopathic formulations, but in reality used it to poison his wife in what became the first known murder using this substance.**

In 1910, Dr Crippen bought hyoscine hydrobromide (scopolamine hydrobromide) from a pharmacy, and a couple of weeks later his wife was seen alive for the last time. Dr Crippen claimed that she had returned to the United States, but when that was found to be untrue a search of the family house revealed human tissue buried in the cellar. Tests determined that the tissue contained high levels of scopolamine, and a scar matched Cora Crippen's medical records. Dr Crippen was convicted of murdering his wife and sentenced to death.

INTERESTING ALKALOIDS

Also within the potato family are plants that produce other types of alkaloids. The piperidine alkaloids, found mainly in tobacco (*Nicotiana* spp.; see pages 98–99), are exploited recreationally and used as insecticides due to their different effects on humans and insects. The capsaicinoid alkaloids in chilli peppers (*Capsicum* spp.; see pages 122–123) use the perception of heat to deter mammals from eating them, but do not affect birds, which disperse their seeds. Lastly, the cestrums (*Cestrum* spp.) contain hepatotoxic kaurene glycosides that are similar to those found in some plants in the daisy family (Asteraceae), such as cocklebur (*Xanthium* spp.) (see pages 162–163).

RIPEN BEFORE EATING

The glycoalkaloid-containing species of *Solanum* are responsible for many cases of suspected accidental poisoning in the United Kingdom. These plants, such as woody nightshade (*S. dulcamara*) and black nightshade, produce berries that children find irresistible. Poisoning is suspected or expected, but symptoms rarely result, except sometimes irritation of the mouth, nausea and possibly vomiting or diarrhoea. The effects are mild because luckily the children who eat these fruit are usually only attracted to the ripe berries – indicated by a change in colour from green to red or black depending on the species. The plant needs the ripe fruit to be eaten in order for the seeds to be dispersed, so the concentration of toxic glycoalkaloids is much reduced. Some species are even considered to be edible when ripe, including the garden huckleberry (*S. scabrum*).

TOP **Woody nightshade (*Solanum dulcamara*) is a climber from Europe and Asia that is also widely introduced elsewhere. Its green fruit ripen through orange to bright red, like a traffic light in reverse.**

False green signals – glycoalkaloids

The potato may seem like one of the most innocuous vegetables, but it has a dark side. There is a reason humans restrict themselves to eating just the tubers of the plant. The leaves, stems and fruit of this herbaceous perennial all contain high levels of toxins known as glycoalkaloids. Even the potato tubers contain some of these toxic compounds, the most common of which are *alpha*-solanine and *alpha*-chaconine. The toxins are there to protect the plant from pests and diseases, and so, in the tubers, they are most concentrated in the peel.

THE ALL-CONQUERING POTATO

PLANT:
Solanum tuberosum L.
COMMON NAMES:
potato
FAMILY:
potato (Solanaceae)
TYPE OF TOXIN:
glycoalkaloids (*alpha*-solanine, *alpha*-chaconine)

SYMPTOMS OF POISONING IN HUMANS:
DIGESTIVE: abdominal cramps, nausea, vomiting, diarrhoea
NEUROLOGICAL: headache, convulsions, paralysis

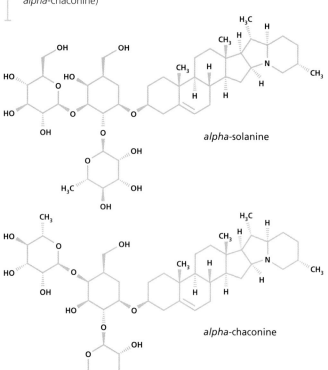

LEFT *Solanum* species contain several glycoalkaloids, these often differing only in the number and types of sugar residues they contain, and how these are connected. Here, the two major compounds found in potato (*S. tuberosum*) are shown.

ABOVE Potato plants (*Solanum tuberosum*) are a common sight on allotments, where they are 'earthed up' at the base to keep light away from the developing tubers. The flowers, with protruding, bright yellow anthers, may be followed by toxic tomato-like fruit.

The potato has a long and glorious history. It was first domesticated from wild varieties approximately 8,000 years ago around Lake Titicaca on the border between modern-day Bolivia and Peru. Potato plants were introduced to Europe by the Spanish in the sixteenth century. In the seventeenth century, European explorers and colonizers spread over the globe, taking the potato with them. Today, the potato is the fourth most important crop in the world, after maize, wheat and rice, and the most important root crop. The plant has been adapted to grow in a wide variety of environments, and the easy storage of the tubers and their nutritional value has supported huge populations at times when other crops have failed.

But it has not been all plain sailing. In 1844, the first signs of a devastating disease appeared in potato crops, late blight or potato blight, caused by an oomycete (*Phytophthora infestans*). At the time, only a few closely related varieties of potato were being grown in Europe and the blight devastated crops from Belgium to Russia. The worst effects were seen in Ireland, where potatoes formed 80 per cent of the population's calorific intake. Between 1845 and 1848, three potato harvests were destroyed, resulting in the deaths of a million people.

THE NASTY SIDE OF THE HUMBLE POTATO

In small amounts, the glycoalkaloids in potatoes contribute to the flavour. However, the amount of these toxic compounds can vary not only with the potato variety, but also the season and growing conditions. Levels can also increase in response to stress. The production of glycoalkaloids is triggered by light, which is why the toxins are most densely concentrated in the leaves and stems of potato plants. It is also the reason why green potato tubers are not good to eat. The green colour that can appear on the tubers is due to the production of chlorophyll for photosynthesis, which is also triggered by light. The presence of green chlorophyll on potatoes is therefore a good indication of higher levels of toxins within. This is a neat way for the potato plant to prevent vulnerable exposed tubers from being eaten by animals or insects.

The not-so-perfect potato crisp

To make a good potato crisp you need a potato with a high starch content and just the right sugars to produce the desired golden colour. In the late 1960s, researchers from the United States Department of Agriculture, Penn State University and the Wise Potato Chip Company thought they had bred the perfect potato for crisps – 'Lenape'. Unfortunately, Lenape potatoes also produce higher levels of glycoalkaloids (around 30 mg per 100 g), even when the plants were not under any environmental stresses. Those who ate them – mostly researchers – developed symptoms much like a stomach bug. The variety was removed from circulation in 1970, just three years after its introduction, but it is still used to breed new varieties, which are tested for their levels of glycoalkaloids before being released.

Commercial varieties of potato contain far less than 20 mg of glycoalkaloids per 100 g of potato (higher levels are associated with a bitter taste) and are completely safe to eat. Stressed potatoes, such as plants affected by blight, eaten by insects, or even bruised and damaged during harvesting and storage, can increase their glycoalkaloid levels as a response. However, the unpleasant, bitter taste of tubers containing high levels of glycoalkaloids usually stops anyone from eating them.

Even elevated levels of toxins from stressed or green potatoes are unlikely to kill, but they can still make you very unwell. Glycoalkaloids damage cells, which will encourage vomiting and diarrhoea, the body's normal response to expel ingested toxins. They may also interfere with chemical messages transmitted between nerves, which can cause headaches and convulsions. Cooking will not destroy these compounds, so it is best to throw green potatoes away.

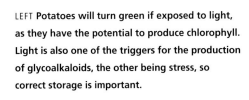

LEFT **Potatoes will turn green if exposed to light, as they have the potential to produce chlorophyll. Light is also one of the triggers for the production of glycoalkaloids, the other being stress, so correct storage is important.**

Amaryllidaceae alkaloids – lycorine

As discussed in Chapters 4 and 10, some species in the amaryllis family contain alkaloids that directly affect the brain. However, out of the 500-plus Amaryllidaceae alkaloids found in the Amaryllidoideae subfamily, it is probably lycorine that is responsible for most poisonings. Lycorine is often present in the highest concentrations and is also the most common Amaryllidaceae alkaloid, occurring in the majority of genera in the Amaryllidoideae subfamily, including the widely recognizable daffodils (*Narcissus* spp.), snowdrops (*Galanthus* spp.), spider lilies (*Lycoris* spp.), crinums (*Crinum* spp.), Natal lilies (*Clivia* spp.), Jersey lily (*Amaryllis belladonna*), blood lilies (*Haemanthus* spp. and *Scadoxus* spp.) and zephyr lilies (*Zephyranthes* spp.).

Easy mistakes

PLANT:
Narcissus pseudonarcissus L., *Lycoris radiata* (L'Hér.) Herb., and others

COMMON NAMES:
N. pseudonarcissus – daffodil, Lent lily; *L. radiata* – red spider lily

FAMILY:
amaryllis (Amaryllidaceae)

TYPE OF TOXIN:
Amaryllidaceae alkaloids (lycorine and several others)

SYMPTOMS OF POISONING IN HUMANS:
DIGESTIVE: nausea, vomiting, diarrhoea

ADDITIONAL SYMPTOMS IN ANIMALS THAT INGEST LARGE AMOUNTS:
NEUROLOGICAL: agitation, seizures, coma
CIRCULATORY: high blood pressure, tachycardia
OTHER: fever

Although severe and lethal poisonings from daffodils in humans have only rarely been reported during the past century or so, the plants do have a reputation for being deadly poisonous, and both pet dogs and cats, as well as cattle, have succumbed after ingestion of the bulbs. In humans, it is well established that daffodil bulbs may cause intensive vomiting and gastric upset even if eaten in only small amounts. As part of a school project in 2009, a group of students aged nine and ten in Suffolk, England, harvested onions (*Allium cepa*) from their own vegetable patch and prepared a soup. After just a short while, several became nauseous, began vomiting and experienced stomach cramps. For 12 students, the symptoms were so severe that they were taken to hospital, although all recovered within a couple of hours and could return home the same day. The reason for their distress was apparently the unintentional inclusion of a single daffodil bulb in their onion soup.

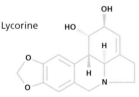

Lycorine

ABOVE **Lycorine is the most widespread alkaloid in the Amaryllidaceae, and can cause severe gastro-intestinal symptoms.**

BELOW **Daffodil bulbs (left) have been confused for onions (right), so should be grown and stored separately. You won't cry if you chop them, but eating them will make you sick.**

LEFT **Daffodils (*Narcissus pseudonarcissus*) are native to Europe, north Africa and temperate Asia, and are popular garden plants and cut flowers that bring welcome colour in early spring.**

A similar confusion, this time between daffodil buds intended as cut flowers and edible spring onions, or chives (*A. schoenoprasum*), happened in 2012 in a community in Bristol, England. A shop displayed the cut daffodils close to the vegetables, and it was assumed by a customer that they were edible greens. In this case, ten people were treated in hospital, and public health authorities warned against the practice of positioning edible vegetables and potentially poisonous cut flowers adjacent to one another in shops. The cut flowers of the alkaloid-containing Amaryllidaceae species pose an additional risk of poisoning due to exudates from the flower stalks contaminating vase water. Although definitely not recommended for drinking, as gastrointestinal problems can result from the presence of bacteria or algae, vase water is usually non-toxic even if the bouquet contains other species of poisonous plants.

Besides alkaloids, daffodils also contain other defensive compounds. The bulbs contain calcium oxalate raphide crystals, similar to those found in members of the arum family (Araceae) (see pages 112–113), which may induce inflammatory responses in animals that eat them raw. The plants also contain lectins, which may add to the gastrointestinal distress caused by the alkaloids (see pages 146–149).

Germination inhibitor

Alkaloids are usually considered to have evolved as protection against herbivores, but in the case of lycorine there is possibly another explanation. It has been shown that this alkaloid has allelopathic activity; the wilting leaves of spider lilies, for example, contain sufficient levels of lycorine to leach into the soil and inhibit the germination of competing plants. Lycorine exerts its effects by inhibiting a plant enzyme in the biosynthetic pathway of ascorbic acid, also known as vitamin C. This vitamin is not only essential for humans, preventing scurvy, but is also vital for the germination of most seeds. However, lycorine and other Amaryllidaceae alkaloids are also active against herbivores and parasites such as eelworms, possibly by inhibiting other enzymes or binding to deoxyribonucleic acid (DNA).

BELOW **Red spider lily (*Lycoris radiata*) contains several Amaryllidaceae alkaloids, including lycorine. Native to eastern Asia, it produces spectacular flowers in late summer.**

Pet's demise

It is not only the alkaloid-containing Amaryllidaceae plants that pose a risk to beloved pets. Ordinary garlic (*Allium sativum*) and onions, enjoyed by humans all over the world, are a lethal risk to dogs and cats even in fairly small amounts. The pungent components in onions are made up of several sulfur compounds that destroy the oxygen-carrying haemoglobin in the red blood cells of dogs and cats, causing acute and potentially lethal anaemia. Thankfully, human haemoglobin is far less sensitive to these *Allium* compounds.

Purging poisons – anthraquinones

Several plants containing anthraquinones are used in herbal medicine for their laxative and purgative properties, particularly rhubarb (*Rheum* spp.) in the dock or buckwheat family (Polygonaceae), aloes such as cape aloe (*Aloe ferox*; see box) in the aloe family (Asphodelaceae), senna (*Senna* spp.) in the legume family (Fabaceae), and the buckthorns (*Rhamnus* and *Frangula* spp.) in the buckthorn family (Rhamnaceae). Used correctly, they can bring relief, but abuse and accidents have unintended consequences.

Rhubarb laxatives

PLANT:
Rheum officinale Baill. and *R. palmatum* L.

COMMON NAMES:
rhubarb, Chinese rhubarb, da huang

FAMILY:
dock or buckwheat (Polygonaceae)

TYPE OF TOXIN:
anthraquinones (glycosides of the anthrones rhein and emodin)

SYMPTOMS OF POISONING IN HUMANS:
DIGESTIVE: vomiting, urgent diarrhoea, ileus (lack of movement)
OTHER: weakness

ABOVE These anthraquinones are found in several plant species in the form of different glycosides. After intestinal bacteria cleave off the sugar residues, they act as powerful laxatives.

From Socotra to the New World

Native to Africa and the Arabian Peninsula, the aloe genus (*Aloe* spp.) contains more than 550 species. Two of these are used globally for their medicinal properties: aloe vera (*A. vera*, syn. *A. barbadensis*), from Saudi Arabia and Yemen; and cape aloe, from South Africa. The name aloe comes from the Arabic *alleoh*, meaning 'shining bitter substance' and alluding to the bitter property of the leaf latex, but the soothing qualities of the leaf gel were also historically valued. So important was aloe that in 333 BC the ancient Greek philosopher and scientist Aristotle persuaded the Macedonian King Alexander the Great to capture the island of Socotra, near the Horn of Africa, in order to control its aloe plantations. And the fifteenth-century

RIGHT Cut leaves of aloe vera (*Aloe vera*) showing the soothing clear gel within.

explorer Christopher Columbus, who had aloe plants on board his ships to heal wounds, once said, 'Four vegetables are indispensable for the well-being of man: wheat, the grape, the olive and aloe.' Dried aloe latex contains bitter compounds, including aloin A and B (barbaloin A and B), the glycosides of aloe-emodin anthrone, an active compound of purgative aloe bitters.

LEFT Garden rhubarb is a robust plant with very large leaves. Shop-bought rhubarb has usually been 'forced' by excluding light, so that the flavour of its pink leaf stalks is less sharp.

USE THE STALKS!

The pink leaf stalks of garden or cultivated rhubarb (cultivars of *Rheum rhabarbarum* and *R. hybridum*) are cooked and used as a fruit to make jams and various desserts. Their malic acid and oxalic acid content gives them a sharpness that is not to everyone's taste, but if cooked and eaten in moderation they are perfectly edible. The large green leaf blade, however, is another matter. It is said that during the First World War the British government encouraged the use of rhubarb leaves as a vegetable, with disastrous consequences. Several deaths attributed to eating cooked rhubarb leaves were reported in the early part of the twentieth century, with symptoms including severe vomiting, abdominal cramp and weakness.

The British government had learnt its lesson by the Second World War, when a Ministry of Information film even advised the public not to include rhubarb leaves in kitchen waste destined to be fed to pigs. It is now accepted that rhubarb leaves should not be eaten. Their higher concentration of oxalates has irritant and corrosive effects on the gastrointestinal tract, with the ingestion of large quantities potentially damaging the kidneys, and the anthraquinones that are present in the fresh leaves are also highly irritant.

Species of rhubarb (*Rheum* spp.) are robust, perennial herbaceous plants. Large, somewhat lobed and toothed leaves arise from a woody rootstock, and a tall, central flowering stem is produced in the summer. The sliced and dried rhizomes and roots of a few species from China are used in traditional Chinese medicine, and since their introduction to Europe in the twelfth to fourteenth centuries they have been incorporated into European herbal medicine.

In dried rhizomes and roots, the anthraquinones are present as glycosides. These are unaffected during their passage through the digestive tract until they reach the colon. Bacterial enzymes in the colon convert these glycosides to the active aglycones (anthrones), principally emodin (6-methyl-1,3,8-trihydroxyanthraquinone), but also rhein, aloe-emodin and chrysophanol. Their effects ease and speed up the movement of food through the colon. This is achieved by enhancing peristalsis, and promoting the colon to secrete water while also inhibiting its absorption of water.

However, if too much of the plant material is taken, or if it has not been prepared adequately (for example, if it has been dried insufficiently) or the wrong part is eaten, symptoms of vomiting, diarrhoea and weakness occur. Conversely, if use of the remedy is continued for more than a few weeks, the stimulating effects can be reversed and ileus can occur – the movement of food through the intestines is slowed or stopped, leading to a build-up and, potentially, a blockage.

RIGHT A leaf and flowerhead of *Rheum palmatum*, which is one of three species whose roots and rhizomes are the source of *da huang*, used in traditional Chinese medicine.

Poisonous proteins – lectins

Unlike venoms produced by animals, plant toxins are rarely proteins. However, the relatively small group of toxic plant proteins that do occur includes some of the most poisonous compounds known. Lectins are a type of protein found in plants and several other organisms, and bind to specific carbohydrates. They perform vital roles – for example, providing protection in plants against pathogens such as fungi.

LEFT The small red and black seeds of the rosary pea (*Abrus precatorius*), a climbing plant grown in many tropical regions, are used in jewellery, rosaries and musical instruments.

BELOW Kidney beans are harvested from the seedpods once they have dried. Using tinned beans is convenient and ensures they have already been cooked sufficiently to make them safe.

Two-fold gut irritation

The most poisonous lectins, such as ricin from the castor oil plant (*Ricinus communis*) in the spurge family (Euphorbiaceae) and abrin from the rosary pea (*Abrus precatorius*) in the legume family, are dimers, consisting of an A-chain and a B-chain, which are linked by a disulfide bond. The B-chain, called a haptomer, is responsible for binding to the surface of a cell, which enables the entire molecule to enter. Once inside, an enzyme in the cell breaks the disulfide bond and separates the two chains. The free A-chain, called an effectomer, is then able to inactivate the ribosomes of the cell, which produce all of its proteins. A single A-chain can inactivate 1,500 ribosomes in just one minute, leading to cell death. The first cells to be affected are those that line the digestive tract, causing severe disturbance of the digestive system. Some of the lectin also passes through these cells and into the circulatory system, which distributes it to the rest of the body, where it can cause more widespread effects such as haemorrhage and multiple organ failure.

Other lectins, such as phytohaemagglutinin in kidney beans (*Phaseolus vulgaris*) in the legume family and curcin from species of *Jatropha* in the spurge family, have only a single haptomer chain that binds to animal cells but lacks the ability to enter them. They are therefore less toxic, mainly causing inflammation and irritation of the gastrointestinal tract. They are thought to bind to, and kill, bacteria in the gut flora, which upsets the gut environment, causing flatulence and diarrhoea.

A RAW DEAL

PLANT:
Phaseolus vulgaris L.
COMMON NAMES:
kidney bean, navy bean
FAMILY:
legume (Fabaceae)

TYPE OF TOXIN:
lectin (phytohaemagglutinin)
SYMPTOMS OF POISONING IN HUMANS:
DIGESTIVE: nausea, vomiting, abdominal pain, diarrhoea

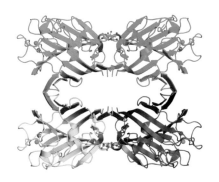

ABOVE **When crystallized, the lectin phytohaemagglutinin forms ring-like superstructures containing four molecules.**

So ubiquitous are lectins, particularly in seeds, that the human diet includes several plants that contain quite high levels of them, such as several types of beans in the legume family. However, problems occur only when the beans are not cooked sufficiently to denature the lectins and make them safe to eat. Denaturing is a process that alters the structure of a protein, and is visible when, for example, you fry an egg.

A number of forms of kidney bean are cultivated. Red kidney beans are the most widely known, but white (cannellini), light speckled and red speckled forms are also grown. Despite the fact that all contain high levels of a mixture of lectins called phytohaemagglutinin (see box), and so need to be soaked and boiled to make them safe, kidney beans are eaten globally and grown in many parts of the world, where they provide a valuable source of dietary protein. They have caused poisoning when children or animals have eaten raw beans or if they have been inadequately prepared. This has happened sporadically, such as when a group of students and their teachers staying in a hostel in the United Kingdom in 1976 ate red kidney beans that had been soaked but not cooked. All of the people who ate more than one or two beans felt sick and vomited, with diarrhoea developing later.

BELOW **The wild relatives of the red kidney bean (*Phaseolus vulgaris*) are native to Central America, and the species was probably first domesticated in Mexico, Guatemala and Peru long before it was taken to Africa in the sixteenth century.**

Blood typing

In 1901, the Austrian physician Karl Landsteiner (1868–1943) described three different types of blood based on the premise that in mixtures of blood from different individuals the red blood cells sometimes clumped together (agglutinated), which destroyed them. He called these groups A, B and O (and today we also recognize the AB type within this system). Similar agglutination had been described already in 1888 in experiments where ricin had been added to blood at the University of Tartu, Estonia. Due to this effect, lectins were originally called phytohaemagglutinins – literally meaning 'plant blood-clumpers' – but this term is today usually used only for the lectins in the kidney bean. Landsteiner showed that lectins from different plants selectively clumped one or some of his blood groups but not others. A lectin from the lima bean (*Phaseolus lunatus*) will agglutinate red blood cells of type A but not B or O, and one isolated from asparagus pea (*Lotus tetragonolobus*) will clump only type O blood. In further studies with different lectins, it has been shown that human blood contains many other types in addition to the ABO grouping, although this categorization is the most important for matching blood in transfusions.

RIGHT **A fresh lima bean (*Phaseolus lunatus*) pod opened to show its green seeds. Mature seeds are called butter beans.**

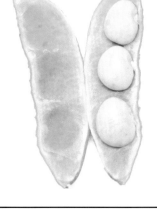

Useful oil and chemical weapon

PLANT:
Ricinus communis L.
COMMON NAMES:
castor oil plant, castor bean, palma christi
FAMILY:
spurge (Euhporbiaceae)
TYPE OF TOXIN:
lectin (ricin), ricinoleic acid
SYMPTOMS OF POISONING IN HUMANS:
CIRCULATORY: tachycardia, circulatory collapse
RESPIRATORY: cyanosis (blue colour to skin), pulmonary oedema
NEUROLOGICAL: stupor, convulsions, coma
DIGESTIVE: burning mouth, nausea, vomiting, stomach cramps, bloody diarrhoea
OTHER: weakness, haemolysis (breakdown of red blood cells), blood in the urine, haemorrhage, multiple organ failure

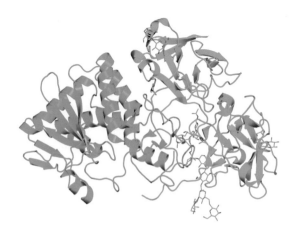

ABOVE RIGHT **Ricin contains a sugar-binding part (blue) and a potentially lethal ribosome-inhibiting part (green).**

BELOW **Castor oil plants (*Ricinus communis*) are grown in countries like Brazil and the seeds harvested for their oil, which is used as a biofuel. The solids left once the oil has been removed are toxic but can be used as a fertilizer.**

The castor oil plant probably originated in east Africa, but due to the global demand for its seed oil an area of more than 1.5 million hectares (5,800 square miles) is now cultivated in tropical, subtropical and warm temperate regions of the world. The castor oil plant is a robust herbaceous perennial that can grow to 5 m (16 ft) or more in height and resemble a small tree, or in cooler conditions it is grown as a single-stemmed annual. The stems and large, lobed leaves are bluish or olive green, or can have a red to bronze colour. The fruit are three-lobed, covered in fleshy bristles and dry on ripening, with each lobe containing a single large seed.

The seeds are decorative and are used in jewellery, as counters in games and in souvenirs for unwary tourists. They are considered inedible and if they are eaten, it is usually by accident or as a result of curiosity; if they are uncooked, this can cause serious poisoning. If seeds are swallowed whole, they pass through the digestive tract without causing symptoms as the hard seed coat ensures the ricin is not released. If they are chewed, however, the ricin can bind to, and enter, the lining of the digestive tract and can pass into the circulatory system, leading to widespread life-threatening effects.

Apart from the toxic lectin ricin, the seeds also contain an oil that is extracted by pressing. The lectin is not soluble in oil, but to make sure the oil produced commercially is free from the toxin, it is heated during the extraction process to denature any traces of the lectin. The oil is possibly best known as a laxative and purgative, and this is also its most ancient use, as described in the Ebers Papyrus, an Egyptian medical text from c. 1552–1534 BC. Seeds have even been found in Egyptian tombs dating from 4000 BC. Due to the presence of a unique compound called ricinoleic acid, a hydroxy fatty acid that makes up approximately 85 per cent of the oil, it has more than 700 industrial uses, including in aviation fuel and biodiesel. It is for this reason that the castor oil plant is cultivated on such a scale.

ABOVE **Seeds of the castor oil plant (*Ricinus communis*) can be purchased to grow this striking ornamental.**

RIGHT **Head of immature green and mature brown castor oil plant (*Ricinus communis*) fruit and seeds, and a bottle of castor oil, which is used both medicinally and industrially.**

Umbrella murder

Although ricin was under development as a chemical warfare agent by a number of countries during both world wars (and possibly afterwards), either as a coating for bullets or shrapnel in the form of a modern arrow poison, or as a component in toxic dust dispersal systems, its use as such has never been confirmed. It did, however, rise to fame in 1978 when the Bulgarian dissident Georgi Markov, who was living in London at the time, was shot with a small pellet containing ricin during his travel to work at the BBC World Service. He noticed a sharp pain in his leg when he was waiting for a bus near Waterloo Station, and turned around to see a man picking up a dropped umbrella. By that evening, Markov had developed severe vomiting and a high fever, and he died in hospital four days later of multiple organ failure. At the autopsy, a small pellet was found embedded in his leg, still containing some ricin, and his death was declared a murder instigated by the Bulgarian secret service.

Ten days earlier, in a parallel event, another Bulgarian defector, Vladimir Kostov, had been shot with an identical pellet in a Paris Metro station. In this case, however, the ricin was not released efficiently from the pellet, which was found under Kostov's skin by his doctors before it could cause serious poisoning.

Bitter cucumbers – cucurbitacins

The Cucurbitaceae or cucumber family, also known as the cucurbits, includes almost a thousand species of plants, many of which are cultivated for their fruits. In fact, the cucumber family has one of the highest proportions of species used for human food, including not only the cucumber (*Cucumis sativus*) but also the watermelon (*Citrullus lanatus*) and the pumpkin and courgette or zucchini (*Cucurbita pepo*), to name but a few. However, some species in this family should not be eaten as they contain cucurbitacins, compounds that can cause anything from stomach cramps to death.

Show-stoppers

PLANT:
Trichosanthes cucumerina L., *Ecballium elaterium* (L.) A. Rich., *Cucumis africanus* L.f. and other species

COMMON NAMES:
T. cucumerina – snake gourd; *E. elaterium* – squirting cucumber; *C. africanus* – wild cucumber

FAMILY:
cucumber (Cucurbitaceae)

TYPE OF TOXIN:
cucurbitacins

SYMPTOMS OF POISONING IN HUMANS:
SKIN: irritation
DIGESTIVE: salivation, stomach cramps, diarrhoea (becoming bloody), sometimes vomiting
OTHER: collapse, kidney failure, liver failure

Cucurbitacin A

LEFT **The cucurbitacins are steroid compounds that act as antifeedant compounds and will cause a bitter taste.**

The cucumber family includes some of the most exotic-looking vegetables known, including the snake gourd (*Trichosanthes cucumerina*), which looks like – you guessed it – a snake. Snake gourds, which can have a twisted appearance, grow up to 2 m (6.5 ft) long and hang from their vines. When the fruit is young and green, it can be safely eaten and is popular in many parts of Asia. But as the fruit matures, it takes on a startling red colour, and at this stage tastes bitter and is dangerous to eat.

Of all the exotic and eccentric cucurbits, the squirting cucumber (*Ecballium elaterium*) has taken things to extremes. Found in Mediterranean regions, the plant has poisonous fruits that are rather unspectacular in appearance, being small, green and covered in bristles. The fruit may not be visually impressive, but it has a trick up its sleeve. When ripe, if it is disturbed or nudged by foraging animals, it explodes away from the stem, squirting a stream of liquid behind it.

ABOVE **Cultivated snake gourds (*Trichosanthes cucumerina*) can weigh up to 1 kg (2.2 lb), and are cooked and eaten when young and green. As they mature to red, they become bitter and inedible.**

A BITTER TASTE IN THE MOUTH

The weird and wonderful collection of fruits produced by the cucumber family contain a huge variety of toxic cucurbitacins, which are noted for their very bitter taste. All cucurbitacin compounds are based on a triterpene cucurbitane skeleton, a steroid-type structure with four fused rings, sometimes with sugars attached (glycosides). The type and quantity of cucurbitacins produced by a plant depends in part on the species and variety. Most cultivated cucurbits have been chosen or bred so that they contain very low levels of the toxins, so it is usually wild species and forms that are dangerous for human consumption. For example, wild cucumber (*Cucumis africanus*) grows in Angola, Namibia, Botswana and South Africa, where a form with large oblong, non-bitter fruits is used as a source of water as well as being eaten as a vegetable. There is, however, another form of wild cucumber that produces small, bitter-tasting fruits that are probably poisonous and unfit for human consumption.

The amount of cucurbitacins in a fruit can also be affected by the growing conditions and can, for example, increase dramatically in response to drought. In 2015, a 79-year-old man in Germany died after eating a courgette (zucchini) that had been given to him by a neighbour. It was thought a recent dry spell had stressed the plant, causing increased levels of toxin to be produced in a fruit that would normally be edible.

LEFT **Wild cucumber (*Cucumis africanus*) is a scrambling plant with 8–10 cm-long (3–4 in) fruits that are covered in short spines, giving rise to one of its other common names, the hedgehog gourd.**

Death in the pot

In the Second Book of Kings in the Old Testament, when Elisha returned to Gilgal (probably in Israel near the border with Jordan), one of his servants gathered a wild vine, *pakku'ot*, which was shredded into a pottage and served to the sons of prophets. On eating it they cried out, 'there is death in the pot!' So Elisha added meal to the pot, after which its contents were no longer harmful.

The name *pakku'ot* translates as 'wild gourds', and the identity of this gourd and the method of detoxification have intrigued researchers. The most likely plant is the colocynth or bitter apple (*Citrullus colocynthis*), although the squirting cucumber has also been suggested. Both have a bitter taste and are drastic purgatives. Inhumane experimentation on dogs in 1919 found that adding a mixture of wheat and corn flour to an infusion of colocynth or squirting cucumber fruit greatly decreased the symptoms experienced by the animals and they made a full recovery.

Stopping cell division – colchicine

The phenethylisoquinolines, of which colchicine is the best known, are almost exclusively found in the autumn crocus family (Colchicaceae), most notably the flame lily (*Gloriosa superba*; see box) and the autumn crocus (*Colchicum autumnale*) itself. The autumn crocus has long been recognized as a treatment for gout, while the flame lily has a number of uses in Ayurvedic medicine. However, the toxicity of colchicine means that great care is required in their use, with severe and fatal poisonings resulting from overdose and abuse, as well as from misidentification of the plants themselves.

DEADLY MISTAKES

PLANT:
Colchicum autumnale L.

COMMON NAMES:
autumn crocus, meadow saffron, naked lady

FAMILY:
autumn crocus (Colchicaceae)

TYPE OF TOXIN:
phenethylisoquinoline (colchicine)

SYMPTOMS OF POISONING IN HUMANS:

CIRCULATORY: hypotension, heart failure

NEUROLOGICAL: weakness, ascending paralysis

DIGESTIVE: thirst, nausea, vomiting, abdominal cramps, diarrhoea

SKIN: alopecia (hair loss) can occur after a delay of several days

OTHER: liver failure, kidney failure, respiratory failure, bone marrow depression

The autumn crocus is native to Europe from Ireland to the Ukraine, and is also cultivated as a garden plant, both in those countries and more widely in temperate regions. It has an unusual growth cycle, with flowers emerging without leaves (naked) from an underground corm in the autumn and dying back by winter. The strap-shaped leaves and fruit capsule develop the following spring, with the capsule maturing and the leaves dying back by the end of the summer. In some populations, the flowers emerge in the spring and the leaves in the autumn. Various parts of the plant are sometimes confused with other species that are edible. Usually it is the leaves that are eaten in the belief that they are wild garlic or ramsons (*Allium ursinum*), but also the flowers are mistaken for saffron crocus (*Crocus sativus*), the corm for 'Japanese ginger' (probably *Zingiber mioga*) and onion, and the capsule for walnut (*Juglans* spp.).

ABOVE **Colchicine, restricted to Colchicaceae, is one of the most toxic alkaloids in plants.**

RIGHT **Autumn crocus (*Colchicum autumnale*) flowers resemble those of a crocus (*Crocus* spp.) but have six stamens, not three.**

Teapot poisoning

The flame lily, or glory lily, could not look more different to the autumn crocus. It is a scrambling climber with spectacular flowers that is native to tropical and southern Africa and tropical Asia. In the areas where it grows naturally, or is cultivated as a source of colchicine, the tubers have been eaten on many occasions either accidentally or for suicide. A case in Sri Lanka in 2016, however, involved the seeds, which contain higher concentrations of colchicine than the tubers. Coriander tea was prepared for the patient by a relative to treat a cold. However, the tea additionally contained flame lily seeds from the farm where the patient worked, which look very similar to coriander (*Coriandrum sativum*) seeds. The patient became extremely unwell but was successfully treated as the seeds were identified quickly, although an extended stay in hospital was required.

ABOVE The flame lily (*Gloriosa superba*) climbs using tendrils at the tips of its leaves. Its spectacular flowers have six flame-like tepals that are swept backwards and six spreading stamens.

Autumn crocus is the most widespread of a genus of about 100–160 species and also has the highest concentrations of colchicine. Colchicine stops cell division (mitosis) by blocking spindle formation by microtubules during the metaphase stage. Its effects are greatest in tissue that has a high turnover of cells, such as the lining of the gastrointestinal tract, and initial symptoms are therefore of digestive disturbance. At toxic levels, colchicine also interferes with other microtubule functions, such as maintenance of cell structure and shape. This can affect all the cells of the body, leading to more serious multiple organ effects over the course of several days, in some cases resulting in death.

USE WITH CARE

The fortunes of autumn crocus as a treatment for gout have varied over time; for a period it was considered heretical, for example. Its use rallied in the early nineteenth century, when colchicine was a key ingredient of *l'eau médicinale d'Husson* patronized by King George IV when he was Prince Regent. Colchicine remained the first line of treatment for gout until the 1960s and 1970s, when it was largely replaced by the development of safer drugs that either prevent gout developing or treat the inflammation caused by the condition. It is, however, still used today for the rare inherited disorder familial Mediterranean fever, in which the sufferer experiences repeated bouts of inflammation in the chest and abdomen.

Colchicine is also being explored and exploited, for use by itself or as a lead compound in the generation of less toxic derivatives, in treatments for cardiovascular disease, diseases of the kidney and cancer. Plant breeders have also found the effects of colchicine on cell division to be a useful tool in the development of polyploid plants, with multiple sets of chromosomes, which can be more resilient than their parents and produce higher yields.

RIGHT The leaves and flowers of the autumn crocus (*Colchicum autumnale*) are not normally seen together, but this drawing shows their comparative sizes as well as the underground corm.

CHAPTER 8

ORGAN FAILURE

The organs of the human body act to maintain a healthy balance in the physiological processes necessary for human life. The liver and kidneys make sure that we get rid of waste and foreign substances, and produce compounds that are necessary for functions elsewhere in the body, such as the blood's ability to clot only when necessary. All these processes can be targets of plant poisons. In this chapter, examples of inhibition, stimulation and tricking of the normal functions of the liver or kidneys by plant compounds are explored.

KILLING THE ORGANS: MECHANISMS OF ACTION

The liver and kidneys are most known for functioning as the body's combined sewers and treatment plant. While Chapter 7 described several poisonous plants that can cause toxicity in the stomach and intestines, even before they are absorbed into the main body, the liver acts as the first line of defence against any substance affecting the body as a whole, and the kidneys are the last contact point before the majority of metabolites leave the body. This makes both these organs more exposed to poisons. With their natural ability to transform and excrete compounds, the liver and kidneys are better protected against low-level exposure to toxic compounds, but on the other hand if they are affected by a poison, the results increase the toxic effects felt throughout the body.

VITAL ORGANS

Both the liver and kidneys produce compounds necessary for vital functions in the rest of the body. The liver is involved in the regulation of blood sugar through the storage of glycogen, which can be transformed into glucose, and by its ability to synthesize glucose from fatty acid or amino acid metabolism during starvation or increased exercise. It also produces most of the proteins that circulate in the blood, including those involved in the blood's ability to clot and prevent fatal blood loss during bleeding. For their part, the kidneys are involved in the production of several hormones, such as those regulating the production of red blood cells, maintaining blood pressure and controlling the concentration and utilization of calcium in the body. These processes are seldom affected by acute poisonings, but in the long run impaired kidney function can lead to secondary effects, adding to the fatalities due to kidney failure from the inherent accumulation of waste products.

LEFT **Cocklebur (*Xanthium strumarium*; see pages 162–163)** is a widespread weed in the daisy family (Asteraceae). Its seeds and seedlings contain atractylosides, which damage cells, particularly of the liver and kidneys, and in addition to poisoning livestock, it has caused the deaths of at least three people in China.

ABOVE **All parts of common comfrey (*Symphytum officinale*) contain pyrrolizidine alkaloids, including the nectar.**

TOXIC EFFECTS

There are many ways in which poisons can disrupt the function of the liver and kidneys, and as this leads to greater exposure to toxins for the rest of the body, it is an effective protective mechanism for plants. In this chapter we encounter the ackee (*Blighia sapida*), which produces a fleshy aril that, if eaten when less than fully ripe, inhibits glucose synthesis in the liver (see pages 158–159). The birdlime thistle (*Carlina gummifera*) and impila (*Callilepis laureola*) are used as emetics and purging medicines, but they actually contain compounds that can disable energy production in cells, and the liver and kidneys are especially vulnerable due to the high concentrations achieved directly after absorption and before excretion (see pages 162–163).

RIGHT **Yellow sweet clover (*Melilotus officinalis*) contains coumarin, which converts to the toxin dicoumarol in mould-spoiled silage or hay.**

In some instances, the metabolic processes that should detoxify poisons instead increase the risk. In the case of pyrrolizidine alkaloids (see pages 164–167), the body's effort to minimize risk instead leads to their transformation into reactive compounds, causing inflammation and an increased risk of cancer. In the kidneys, a similar situation occurs with aristolochic acids present in birthworts (*Aristolochia* spp.; see pages 168–169). These compounds destroy the kidney cells and can cause cancer, possibly after activation by metabolic enzymes involved in the kidneys' regulation of the calcium balance of the body.

In other instances, plant compounds inhibit the function of liver enzymes. Coumarins (see pages 170–171) can obstruct the function of vitamin K, which is necessary for the production of proteins responsible for blood clotting. Members of the poppy family (Papaveraceae; see pages 172–173) are especially rich in some alkaloids that can inhibit the enzymes responsible for the detoxifying capability of the liver, and so get distributed to the rest of the body intact. In the case of sanguinarine, this leads to an effect on the blood vessels, which start to leak fluids, causing oedemas and impairing function in several organ systems (see pages 174–175).

Soapberries – hypoglycins

Ackee is a forest tree from west Africa that can reach 15–24 m (50–80 ft) in height. It was introduced to Jamaica, probably on a slave ship, from where it was collected and taken to England in 1793 by Captain William Bligh, after whom it gets its genus name, *Blighia*. Bligh was visiting Jamaica as captain of HMS *Providence* on his second attempt to introduce breadfruit (*Artocarpus altilis*) to the West Indies from Tahiti, his first attempt in command of the HMS *Bounty* having famously ended with the mutiny of his crew.

RIPE FOR THE PICKING

PLANT:
Blighia sapida K.D.Koenig

COMMON NAMES:
ackee

FAMILY:
soapberry (Sapindaceae)

TYPE OF TOXIN:
hypoglycin A

SYMPTOMS OF POISONING IN HUMANS:
NEUROLOGICAL: headache, paraesthesia (pins and needles), convulsions, altered consciousness, coma
DIGESTIVE: vomiting
OTHER: hypoglycaemia, hypotonia (decreased muscle tone), generalized weakness or lethargy, hypothermia

Although it has also been introduced to the other Caribbean islands, Central America and Florida, ackee is widely eaten only on Jamaica. In fact, it is Jamaica's national fruit, and ackee and saltfish is the national dish. The leathery fruit are 7.5–10 cm (3–4 in) long, bright red or yellow-orange when ripe, and split open into three sections to expose three shiny black seeds, each surrounded by a large yellow or whitish aril. Only arils from ripe fruit that have naturally split open are eaten. To remove any residual toxicity, they are cleaned of all red fibre (the aril membrane) and boiled, and the water they are boiled in is discarded. Cooking unripe arils does not destroy their toxicity.

BELOW **The ackee tree (*Blighia sapida*) has pairs of glossy leaves. Its fruit ripen to red and, when they split open, the cream arils within can be eaten after cooking.**

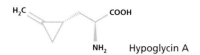

ABOVE **The unusual amino acid hypoglycin A is not incorporated into proteins. In humans, it causes dangerously low blood sugar levels by preventing the body from making glucose.**

Jamaican vomiting sickness

Before the toxicity of ackee was understood, eating unripe arils frequently caused poisoning known as Jamaican vomiting sickness, which occurred as an annual epidemic. Symptoms included vomiting, convulsions and, frequently, also coma and death, with mortalities being more common in children, particularly those already suffering from malnutrition. The underlying cause was eventually linked to the consumption of unripe ackee arils. This results in low blood sugar levels (hypoglycaemia) through a blockade of the liver's ability to synthesize glucose and a reduction in fatty acid metabolism (both normal routes for increasing levels of blood sugar), as well as depletion of the liver's carbohydrate reserves.

Poisoning is due to the presence of an amino acid derivative, hypoglycin A (2-amino-3-(methylenecyclopropyl)-propionic acid), which is also found in other plants of the soapberry family, such as lychee (*Litchi chinensis*; see box). In ackee, the concentration of hypoglycin A is high in unripe arils and reduces significantly as they ripen, although low levels remain in the aril membrane. The seeds also contain the less toxic hypoglycin B (the *gamma*-glutamyl conjugate of hypoglycin A), with concentrations significantly increasing as the seeds ripen.

Butter and cheese

Two types of aril are recognized in Jamaica: 'butter', which is soft and yellow and said to have a better taste; and 'cheese', which is hard and cream-coloured, and is preferred by the canning industry. Tinned ackee arils are sold around the world, primarily to Jamaicans living abroad. Their import was banned in the United States in 1973 due to toxicity concerns, but is now permitted if controls are in place that ensure products contain only properly ripened and processed ackee arils.

Poisonings have decreased in Jamaica but they still occur sporadically there and elsewhere. For example, in 2000 in Haiti, when severe floods destroyed crops, many cases resulted from people eating unripe fruit, including 50 fatalities. Around the same time, 16 children in Suriname also died, most likely after witch doctors used ackee fruit to 'cure' them of acute diarrhoea and other diseases. There were no new deaths after the community was informed of the toxicity of these fruit.

Lychees

Since the early 1990s there have been a number of outbreaks of hypoglycaemic encephalopathy (brain disease) in India, Bangladesh and Vietnam, affecting mainly children and with a high mortality rate. These outbreaks were found to coincide with the lychee harvest, and once exposure to an infectious agent or a pesticide had been ruled out, the possibility that these edible fruit might be the cause was considered. At the time, lychee seeds and arils were known to contain a compound related to hypogylcin A called methylenecyclopropylglycine (MCPG). More recently, analysis of fresh lychee arils found that they also contain hypoglycin A itself, with concentrations of this and MCPG being higher in unripe arils. It was confirmed that all the children involved had eaten lychee and, as with ackee poisoning, malnourished children were the most seriously affected.

TOP **Lychee (*Litchi chinensis*) is an evergreen tree from tropical Asia and southeast China. Its fruit have a thin, rough skin around a sweet, aromatic white aril that encloses a single elongated seed.**

ASTERACEAE

The Asteraceae, known as the daisy, sunflower or aster family, is one of the largest plant families, with over 25,000 species in more than 1,600 genera. The classical name for this family, the Compositae, alludes to the composite structure of the flowerhead, the capitulum, which makes most members of this cosmopolitan family quite easy to spot. What we think of as a flower is actually a flowerhead made up of a number of small (even tiny) flowers, the florets. There are two main floret forms: the 'disc' floret, which is tubular; and the 'ray' floret, which is elongated to resemble a single petal. In the daisy (*Bellis perennis*) and sunflower (*Helianthus annuus*), a central 'disc' of disc florets is surrounded by ray florets, while in the dandelion (*Taraxacum officinale*), the flowerhead is made up entirely of ray florets.

ABOVE Roman chamomile (*Chamaemelum nobile*) is a low-growing perennial herb from western Europe and north Africa. Tea is more usually made from flowers of German chamomile (*Matricaria chamomilla*).

JERUSALEM ARTICHOKE TO ROMAN CHAMOMILE

Many members of this large family are grown as garden and cut flowers, while others provide food or medicine. The kernel (the true seed) of the sunflower 'seed' (the actual fruit) is commonly eaten, and a widely used cooking oil is extracted from it. Culinary oil is also extracted from other species, including the safflower (*Carthamus tinctorius*). In addition, the leaves of lettuce (*Lactuca sativa* and other species), chicory (*Cichorium intybus*) and tarragon (*Artemisia dracunculus*), the flowerheads of the globe artichoke (*Cynara cardunculus*), and the tubers of the Jerusalem artichoke (*Helianthus tuberosus*) and some dahlias (*Dahlia* spp.) are also eaten.

For the size of this evolutionarily successful family and the range of compounds produced by its members, there are surprisingly few plants that are used medicinally. In fact, Asteraceae is considered to be medicinally poor. Some species, however, are very widely used, such as German chamomile (*Matricaria chamomilla*, syn. *M. recutita*) and Roman chamomile (*Chamaemelum nobile*), used to make calming chamomile tea, and the immunity-boosting coneflower (principally *Echinacea purpurea*).

BELOW Dandelion (*Taraxacum officinale*) is a widespread perennial with toothed leaves and a flowerhead formed from ray florets.

ECOLOGICAL IMPORTANCE

We may not use many Asteraceae plants as foods or medicines, but we do rely on them for their contribution to the diversity and stability of grassland, shrubland and woodland in most areas of the world, the exceptions being Antarctica and tropical rainforests. Very few are trees or shrubs, with some 98 per cent being herbaceous. This family often dominates the vegetation that reclaims cleared land, with some of these pioneer species being noxious weeds, including common ragwort (*Jacobaea vulgaris*; see pages 166–167). While common ragwort may contain pyrrolizidine alkaloids, which are highly toxic to farm animals, it is also a vital food plant for many insects. In addition, the fruit of a number of species are an important source of food for birds and small mammals, including nyjer or niger (*Guizotia abyssinica*) and sunflower, which are sold for this purpose.

ABOVE **Many birds feed on the seeds of plants in the daisy family, such as this female American goldfinch (*Spinus tristis*) clinging upside down to a bull thistle (*Cirsium vulgare*) to dig for seeds.**

BELOW **Milkthistle (*Silybum marianum*) has impressive heads of disc florets and the enclosing bracts have long spines. Its large leaves are marbled with white and also have spines along their toothed edges.**

LIVER TOXINS

Several Asteraceae species that contain pyrrolizidine alkaloids (see pages 164–167) – including coltsfoot (*Tussilago farfara*), hemp agrimony (*Eupatorium cannabinum*), Joe Pye weed (*E. purpureum*), butterburs (*Petasites* spp.) and several species of ragwort and groundsel (*Senecio* spp.) – are used medicinally, but this is now discouraged due to concerns about potential liver toxicity. Interestingly, the fruit of another family member, milkthistle (*Silybum marianum*), are used in herbal medicine as a liver tonic and silymarin, the mixture of compounds isolated from the fruit, is used to treat poisoning by deadly hepatotoxic fungi such as death cap (*Amanita phalloides*) and destroying angel (*A. virosa*).

Other toxins found in the daisy family are terpenoid compounds, including the atractylosides, which are diterpene glycosides. Found in the birdlime thistle and several other plants, atractylosides are toxic to the cells of the liver and kidneys (see pages 162–163). The family also contains an insecticide originally isolated from the pyrethrum daisy (*Tanacetum coccineum* and *T. cinerariifolium*), previously widely used due to its low toxicity in humans (see page 217). A more unusual toxin, called tremetol, is found in white snakeroot (*Ageratina altissima*). It includes tremetone, a benzofuran ketone, and causes trembles in livestock and milk sickness in humans (see page 191).

Running low – atractylosides

The cardiotoxic aconite alkaloids that are derived from the terpenoid precursor geranylgeranyl diphosphate are discussed in Chapter 3. Here, we look at the similarly derived terpene compound atractyloside, which lacks nitrogen atoms, and instead of producing heart arrhythmias causes severe injury to liver and kidney cells.

Atractyloside

ABOVE **Atractyloside is a kaurene-type diterpene glycoside, which has an unusual presence of sulfate functions on the sugar moiety. It is an inhibitor of energy production in cells.**

PUSHING UP THE DAISIES

PLANT:
Carlina gummifera (L.) Less. (syn. *Atractylis gummifera* L.), *Callilepis laureola* DC., *Xanthium strumarium* L., and others

COMMON NAMES:
C. gummifera – birdlime thistle, glue thistle, stemless atractylis; *C. laureola* – impila, oxeye daisy; *X. strumarium* – cocklebur

FAMILY:
daisy (Asteraceae)

TYPE OF TOXIN:
kaurene glycosides (atractyloside, carboxyatractyloside)

SYMPTOMS OF POISONING IN HUMANS:
DIGESTIVE: abdominal pain, vomiting, diarrhoea
NEUROLOGICAL: headache, dizziness, seizures, coma
OTHER: low blood sugar (hypoglycaemia), liver and renal failure

A number of plants, primarily from the daisy family, contain the diterpene glycoside atractyloside and its derivate carboxyatractyloside, which deplete energy production in mitochondria, effectively killing affected cells if the inhibition is sustained for more than a brief time. As the liver can be considered a first line of defence in humans, its cells are especially susceptible to this poison. With larger ingestions, the cells of the kidney – responsible for excreting atractylosides – will also be affected.

Throughout the Mediterranean, the birdlime thistle (*Carlina gummifera*) is known as a lethally poisonous plant. The powdered root of the species was apparently mixed into couscous and used as a homicidal poison, and in Morocco it is traditionally used as an emetic. The plant is often confused with 'wild artichokes' (possibly *Cynara humilis*) and the root has a sweetish taste, thus posing a particular risk for children. More than 200 deaths related to birdlime thistle have been reported since the middle of the nineteenth century.

LEFT **Birdlime thistle (*Carlina gummifera*) has a basal rosette of spiny leaves that often die back early. Pollinators are attracted with a glue-like scent, hence its other name of glue thistle.**

HEALTHY OR NOT?

In South Africa, it has been recognized since the 1970s that the medicinal plant impila (*Callilepis laureola*) causes liver failure. In the Zulu language, *impila* means 'health', and the plant has been used by this ethnic group to treat stomach complaints, coughs and tapeworm infestations, as well as to ward off evil spirits. The mortality associated with the species is estimated to be high due to the rapid progression of the poisoning. Based on early reports, death usually occurs less than 24 hours after the first symptoms appear, and many poisoned patients never even reach hospital. As liver and kidney failure is the main result of exposure to the toxin, mortality is high even with hospital treatment, especially in children and malnourished patients.

The cocklebur (*Xanthium strumarium*) is used in traditional Chinese medicine 'to dispel wind-cold headache and wind-damp generalized stiffness and pain', which modern herbalists often translate as sinusitis, nasal congestion and allergies with respiratory symptoms. Even though there are reports of adverse reactions such as vomiting and muscle spasms after the use of the herbal remedy, cocklebur is perhaps most often associated with animal toxicity.

ANIMAL RISKS

Cocklebur, which has become a cosmopolitan weed, is known to cause poisonings in cattle, sheep and pigs. The seeds and seedlings are the most toxic parts, with mature plants containing very low levels of atractylosides before seed production. Cocklebur probably originated in North America, and several other plants across the Americas containing atractylosides cause poisonings in domesticated and farmed animals. For example, the creeping oxeye daisy (*Pascalia glauca*, syn. *Wedelia glauca*) was responsible for a small cluster of lethal intoxications in an Argentinian zoo in 2010, where it contaminated alfalfa (*Medicago sativa*) hay and killed seven axis deer (*Axis axis*) and one llama (*Lama glama*).

In South America, the green cestrum (*Cestrum parqui*), in the potato family (Solanaceae; see page 139), contains the compound parquin, a kaurene glycoside, which differs from atractyloside only in its attached sugars. It has caused cattle losses on farms in the areas where it is native. Green cestrum has also been introduced as a garden and windbreak plant in South Africa, although poisonings there are practically unheard of. However, the related inkberry or coerana (*C. laevigatum*) was identified as the cause of Chase Valley disease in cattle, named after the location in KwaZulu-Natal of several occurrences of the disease in the early twentieth century. It is not clear if this species also contains toxic diterpene glycosides.

LEFT The oval-shaped fruits of cocklebur (*Xanthium strumarium*) each contain two seeds. They are covered in hooks that easily attach to fur and clothes, enabling the plant to spread around the world.

BELOW Green cestrum (*Cestrum parqui*) is a medium shrub. It bears heads of yellow flowers that are fragrant in the evening and are followed by purple-brown to black berries.

Famine-food poisoning

One instance of a recent mass human poisoning occurred after destructive monsoon flooding in Bangladesh in the autumn of 2007, when 76 cases of vomiting, liver damage and altered mental state were identified. The mortality rate in the outbreak was 25 per cent, mostly in children under the age of 15. The explanation turned out to be ingestion of *ghagra shak*, the local name for cocklebur, as a famine food following the destruction of local rice crops.

Hepatic harm – pyrrolizidine poisoning

Pyrrolizidine alkaloids are found in a wide variety of plants, mainly belonging to the unrelated borage (Boraginaceae), legume (Fabaceae) and daisy families. The pyrrolizidines are produced by the plants as a deterrent against herbivores that might eat them. But these compounds do not just taste bad – they can eventually be fatal.

The toxicity of common ragwort, a member of the daisy family, to cattle and horses is fairly well known, even if the chemical culprit within the plant is not. Pyrrolizidine alkaloids can also find their way into human food in several ways, from contamination of cereal crops, to inclusion in herbal teas and remedies, and in tainted honey. Some are more worrying than others – for example, exposure from honey is usually negligible but might be a risk for individuals who consume large amounts of honey containing high levels of these toxic alkaloids.

Pyrrolizidines in detail

Pyrrolizidine alkaloids are a family of compounds that are almost as diverse as the plants that contain them. They are named for their common chemical feature, a pyrrolizidine unit of two fused five-membered rings with a central nitrogen atom. A huge number of chemical variants are possible, not all of them toxic, by substituting different groups around the rings. These alkaloids are poisonous only after metabolic activation transforms them into reactive compounds, which cause inflammation and damage to deoxyribonucleic acid (DNA). Most damage is caused to the veins draining blood away from the liver and back to the heart and lungs. The veins become blocked (veno-occlusive disease), and the blood that does escape can carry the toxins to the lungs and other organs. After a single ingestion, the body is usually capable of reversing the damage and patients suffer only abdominal pain, vomiting and diarrhoea. However, if the exposure is massive or extended over time, the liver will be irreversibly damaged, finally leading to fatal liver failure. The worst aspect of pyrrolizidine poisoning is that there is no antidote. The best that can be done is to treat symptoms and prevent exposure to the toxins in the first place.

LEFT **Common comfrey (*Symphytum officinale*) is widespread in Europe and introduced elsewhere. It is a vigorous plant with fleshy roots, large basal leaves and upright, branched, leafy flowering stems.**

Toxic tea?

PLANT:
Symphytum officinale L.
COMMON NAMES:
common comfrey, knitbone, boneset
FAMILY:
borage (Boraginaceae)
TYPE OF TOXIN:
pyrrolizidine alkaloids

SYMPTOMS OF POISONING IN HUMANS:
After a single ingestion:
DIGESTIVE: abdominal pain and vomiting
After repeated doses:
LIVER: swollen abdomen, liver disease, jaundice
LUNGS: accumulation of fluid, restricting breathing

Symphytine

ABOVE **Symphytine is one of the pyrrolizidine alkaloids that contain the double bond in the core structure that is necessary for metabolic activation.**

Common comfrey (*Symphytum officinale*) has been used in traditional herbal medicine for more than 2,000 years, and many of its common names – such as knitbone and boneset – refer to this long history of medical application. Even the name of its genus, *Symphytum*, comes from the ancient Greek words *sympho*, meaning 'to bring together', and *phyton*, meaning 'plant'. Applied in external poultices, comfrey is said to aid wound healing and have anti-inflammatory effects. The plant has also been used internally, to treat gastrointestinal disorders, gout and a host of other conditions, but with less sympathetic results.

The use of species of comfrey (*Symphytum* spp.) in herbal teas has raised concerns in recent times that the pyrrolizidine alkaloids they contain may cause liver damage. Indeed, several cases of poisoning have been documented, from Ecuador to Egypt, and from the United States to Hong Kong. Comfrey teas are usually made from the leaves, which contain lower levels of the alkaloids than the roots, although the concentrations can vary with the age of the plant and the species. For example, rough comfrey (*S. asperum*) contains much higher concentrations of pyrrolizidines than common comfrey, but the two species are easily confused. A hybrid between them, Russian comfrey (*S. × uplandicum*), also contains high levels of pyrrolizidine alkaloids. Because of the potential for harm, several countries have banned the use of comfrey in herbal teas and remedies, except for those used as topical applications.

Kill or cure

Several pyrrolizidine alkaloid-containing plants are used in traditional herbal remedies around the globe, sometimes with fatal consequences. In ancient Greece, an aphrodisiac named *satyrion* was used to treat the love-sick. As it was made from common ragwort, it may well have made people feel sick, but not due to love. In the borage family, hound's tongue (*Cynoglossum* spp.), fiddleneck (*Amsinckia* spp.) and viper's bugloss (*Echium* spp.) are among the most commonly used medicinal plants that are particularly known for their toxicity. In India, three cases of veno-occlusive disease were documented from the use of a species of heliotrope known as hathisunda (*Heliotropium ellipticum*, syn. *H. eichwaldii*). The preparations had been prescribed to treat epilepsy and vitiligo but resulted in severe liver damage, and at least two of the three patients died as a direct result (the outcome of one case is not known).

TOP **Common viper's bugloss (*Echium vulgare*), a blue-flowered perennial herb, is native to Europe and parts of temperate Asia, and introduced elsewhere.**

BAD BREAD

Contamination of cereal crops by pyrrolizidine-containing plants is a real threat, particularly in arid climates where these species often thrive, and there have been several cases of mass poisoning involving contaminated bread. Following a dry spell there may be a much higher proportion of the toxic weeds growing among the crop. The threshing process, to loosen the grain from the crop, will also loosen the seeds from any plants contaminating the harvested cereals. The following stage, winnowing, removes the chaff and other undesirable contaminants from the grain using a current of air. If this is not done properly, toxic seeds can remain and may be processed into the flour.

The results of such contamination have been a sudden outbreak of veno-occlusive disease, which can often have a high death toll. The disease was first reported in South Africa in the 1920s, when it was linked to contamination of cereals with ragwort species (*Senecio ilicifolius* and *S. burchelli*). In India in the 1970s, an outbreak was linked to cereal contamination with rattlepod (*Crotalaria* sp., in the legume family). However, the most extreme case occurred in a region of northwestern Afghanistan in the 1970s. Two years of low rainfall meant that a higher proportion of charmac (*Heliotropium popovii* ssp. *gillianum*, in the borage family) plants were growing among grain crops. An estimated 7,900 people (of a total population of 35,000) were affected, of whom around 2,000 died. The toxicity of the plants was compounded by poor nutritional health in the region, making the local population more susceptible to the effects of the toxic alkaloids.

POISONOUS PASTURES

PLANT:
Jacobaea vulgaris Gaertn. (syn. *Senecio jacobaea* L.)

COMMON NAMES:
common ragwort, stinking willie, tansy ragwort, benweed

FAMILY:
daisy (Asteraceae)

TYPE OF TOXIN:
pyrrolizidine alkaloids (jacobine, retrorsine, seneciphylline, monocrotaline and others)

SYMPTOMS OF POISONING IN HUMANS:
as for *Symphytum officinale* (see page 165)

SYMPTOMS OF POISONING IN ANIMALS AFTER REPEATED INGESTION:
NEUROLOGICAL: apathy, uncoordinated walking
DIGESTIVE: loss of appetite, diarrhoea
OTHER: phototoxicity induced by liver failure

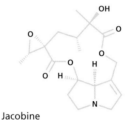
Jacobine

LEFT The acid esterified to the pyrrolizidine core of jacobine is a so-called necic acid. Several of these are known and do not affect the toxicity of the alkaloid.

BELOW Although cows and other livestock find ragwort (*Jacobaea vulgaris*) unpalatable, they are at risk from it if they are allowed to graze on pasture in which other plants become scarce.

ABOVE **The yellow flowers of ragwort (*Jacobaea vulgaris*) are a nectar source for many insects, including gatekeeper (*Pyronia tithonus*) and meadow brown (*Maniola jurtina*) butterflies.**

The bright yellow daisy-like flowers of common ragwort are a common sight in much of the world, and the plant is often seen growing along roadsides and on uncultivated land. Together with many related plants, such as South African ragwort (*Senecio inaequidens*) and groundsel (*S. vulgaris*), common ragwort is well known as a potential hazard to cattle, horses, pigs and deer, and is listed in several countries – including the United States – as a 'noxious weed'. The plant contains high levels of toxic pyrrolizidine alkaloids, which can cause fatal liver damage in animals that eat it. Sheep and goats seem to have a much higher tolerance to these toxins, perhaps because their liver metabolizes them in a different way. Most horses and cows do not eat the bitter-tasting fresh plant unless there is a shortage of food. The biggest concern is that the plant finds its way into hay, as dried ragworts lack the bitter warning taste and animals will feed on them unawares. Farmers and horse owners should be careful to remove ragwort from grazing areas.

The positive side of poisonous ragwort

A single common ragwort plant can produce thousands of yellow flowers and tens of thousands of seeds in a single season. With the obvious danger the species poses to cattle and other farm animals, it may seem like a straightforward argument to eradicate it from as many places as possible so that the wind-dispersed seeds do not find their way back into pastures. But ragworts are not all bad. In the United Kingdom, these plants provide food for at least 77 species of foliage-eating insects. Thirty insects feed exclusively on the plant and eight are listed as 'nationally scarce'. The plants also provide nectar for 170 insect species, often solitary bees and butterflies. So these noxious plants are in fact vital in maintaining biodiversity.

Deadly diets – aristolochic acid nephropathy

The birthwort family (Aristolochiaceae) consists of around 500 species, some of which are grown for their spectacular or unusual flowers. As the common name birthwort suggests, these plants also have a role in traditional medicine. This stretches back to the time of the ancient Greeks – indeed, the name *Aristolochia* comes from the Greek *aristos*, meaning 'best', and *lochia*, meaning 'childbirth'. This medical link was made according to the doctrine of signatures (see box, page 127), and is based on the shape of some flowers, which can resemble the birth canal, rather than any proven efficacy. Plants in the birthwort family grow around the globe and their use in systems of traditional medicine has not been restricted to Europe. They are included in Indian and Chinese herbal remedies, for example, for conditions as varied as oedema and arthritis. The presence of the highly toxic aristolochic acid in many of these plant species has doubtless led to numerous deaths over the centuries of their use.

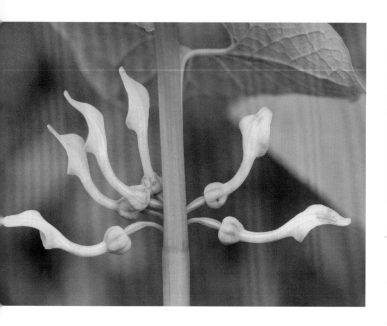

ABOVE **The European birthwort (*Aristolochia clematitis*) is a perennial herb with heart-shaped leaves. Pipe-shaped flowers arise in groups of 1–8 at the leaf axils.**

RIGHT **The aristolochic acids, here aristolochic acid I, are carcinogenic and contain a nitro group unusual for plants.**

Aristolochic acid I

Mysterious Balkan kidney disease

PLANT:
Aristolochia clematitis L. and
A. fangchi Y.C.Wu ex
L.D.Chow & S.M.Hwang

COMMON NAMES:
A. clematitis – European birthwort, Dutchman's pipe, pipevine; *A. fangchi* – guang fang ji, Chinese snakeroot

FAMILY:
birthwort (Aristolochiaceae)

TYPE OF TOXIN:
aristolochic acid

SYMPTOMS OF POISONING IN HUMANS AFTER REPEATED INGESTION:

KIDNEY: progressive renal disease, kidney failure, cancer
OTHER: fatigue, anaemia

In 1956, a study was made into a strange occurrence of kidney disease among people in farming villages in Bulgaria. Further cases were found in communities along tributaries of the River Danube in Croatia, Serbia, Bosnia and Romania. Symptoms of anaemia, weakness and coppery-coloured skin often progressed to kidney failure, with patients requiring dialysis or kidney transplant. The disease was named Balkan endemic nephropathy (BEN) and affected 25,000 people.

Genetic, environmental and viral causes were all rejected, and the true cause of the disease remained a mystery for decades. Then, in 1991, when a species of *Aristolochia* was implicated in cases of kidney failure in Belgium (see box), scientists looked at the Balkan cases again. Examining the

ABOVE **The Manchurian birthwort (*Aristolochia manshuriensis*), from China and Korea, is the source of the traditional Chinese medicine *guan mu tong*, but its use has been discontinued since the serious effects of aristolochic acid have become known.**

diets of those affected, they uncovered a tradition of baking bread from locally grown wheat, and found European birthwort (*A. clematitis*) growing among the crop. Seeds from the birthwort plants had found their way into the flour and poisoned the bread. Examination of samples from patients showed tell-tale signs of kidney damage due to aristolochic acid.

ARISTOLOCHIC ACID

Aristolochic acid is extremely toxic to the kidneys and a potent carcinogen. It causes severe and permanent kidney damage, as well as tumours in the kidneys and urinary tract, although it is unclear whether the same mechanism is responsible for the two different clinical results. What is known is that aristolochic acid is metabolized in the body into a derivate that binds strongly to DNA. This interaction with DNA causes specific mutations (the adenine–thymine base pairs are inverted to thymine–adenine) that are characteristic of this type of poisoning and can be used to confirm cases. The more aristolochic acid to which an individual has been exposed, the greater the damage caused and the greater the severity of the disease. Symptoms can be delayed for weeks or even longer after exposure has stopped, as unaffected parts of the kidneys can at least initially compensate for the damage. However, eventually such compensation diminishes and the kidney disease inevitably progresses.

The wrong fang ji

In 1991, a number of patients – mostly women – appeared with kidney failure in medical centres and hospitals in Belgium. The cases were linked by the fact that the individuals had taken a herbal remedy as part of a Belgian weight-loss programme. It was found that the pills they had been given contained guang fang ji (*Aristolochia fangchi*), a source of aristolochic acid. Han fang ji (*Stephania tetrandra*), in the moonseed family (Menispermaceae), had been the intended ingredient, but the pinyin names (transliterations of the Chinese names) for the drugs had been confused. This simple case of mistaken identity affected more than a hundred people, many of whom required dialysis or transplant, and at least three of whom died. As a result of the incident in Belgium, the sale of remedies containing *Aristolochia* species has been restricted in several European countries. In the United States, the Food and Drug Administration has issued alerts about the dangers of aristolochic acid. Unfortunately, however, this is not the end of the story, as cases of kidney failure and cancers linked to the herbal use of *Aristolochia* species continue to occur.

Thinning the blood – coumarins

Coumarin was first isolated from tonka beans (*Dipteryx odorata*, in the legume family) and named after *coumarou*, the French word for the plant. Since then, more than 1,300 coumarins have been identified in plants from over 100 families, as well as in fungi and bacteria. Plants that contain coumarins are used in herbal medicine, and as flavourings and scents (see box).

LEFT Most species of sweet clover are native to Europe and Asia, but have naturalized widely elsewhere. Those with high levels of coumarin have caused haemorrhagic disease in cattle.

Mouldy cattle feed

PLANT:
Melilotus officinalis (L.) Pall. and *M. albus* Medik.

COMMON NAMES:
M. officinalis – sweet clover, yellow sweet clover, melilot;
M. albus – white sweet clover, white melilot

FAMILY:
legume (Fabaceae)

TYPE OF TOXIN:
benzopyrone (coumarin)

SYMPTOMS OF POISONING IN LIVESTOCK:

CIRCULATORY: subcutaneous haemorrhage (bleeding under the skin), spontaneous bleeding

NEUROLOGICAL: weakness, inability to move

Coumarin

LEFT Coumarin is considered to have no or very low toxicity, but in its dimeric form it acts as a blood thinner.

When North American farmers introduced sweet clover species (*Melilotus officinalis* and *M. albus*) from Europe in the early twentieth century as cattle feed, they did not realize that they were also importing a potential source of poisoning. Cases of a previously unseen haemorrhagic disease were soon occurring in livestock with devastating effects, ruining the livelihoods of many farmers. The source was eventually traced to the sweet clover. Fresh sweet clover was not a problem – in fact, the cattle did not like to eat it, as although it smelt sweet it had a bitter taste. The problem arose when sweet clover was fed to cattle in the

The smell of newly mown hay

Coumarin gives sweet clovers their fragrance and is also responsible for the sweet smell of newly mown grass. It has been described as 'sweet, aromatic, a creamy vanilla bean odour with nut-like tones that are heavy, but not sharp or brilliant'. Many plants that contain coumarin have been used for their smell, including sweet woodruff (*Galium odoratum*, syn. *Asperula odorata*) in the coffee family (Rubiaceae), which was used as a strewing herb and moth repellent, and also to flavour *Maiwein* (may wine) in Germany. Sweet grass (*Anthoxanthum nitens*, syn. *Hierochloe odorata*), in the grass family (Poaceae), is used to flavour Żubrówka, a vodka traditionally made in eastern Europe. It is also known as holy grass as it was used to scent churches. Since 1954, when concerns about the potential toxicity of coumarin were raised, there have been limits on the levels allowed in food and drinks. Chinese cinnamon (*Cinnamomum cassia*), in the laurel family (Lauraceae) and the most common commercial source of the spice cinnamon, is now thought to be the source of the most coumarin in diets.

BELOW Sweet woodruff (*Galium odoratum*), from Europe and temperate Asia, is a low-growing, sweetly scented perennial with whorls of soft leaves and clusters of white flowers.

BELOW These Chinese cinnamon (*Cinnamomum cassia*) sticks are a single thick layer of bark, whereas those of true or Ceylon cinnamon (*C. verum*) have several soft, thin layers.

form of hay or silage, which when spoiled due to mould caused spontaneous and uncontrollable bleeding.

It took many years and several important chance encounters for the active compound in spoiled sweet clover to be identified. Spoiled silage made from *Melilotus dentatus*, a species of sweet clover with a low coumarin content, was found not to cause haemorrhagic disease in cattle, whereas the addition of coumarin to alfalfa hay that was then allowed to spoil did cause the disease. Furthermore, it was found that the coumarin itself was inactive, but that when sweet clover spoils under the influence of *Aspergillus* mould, the compound is converted to the active dicoumarol, which is an anticoagulant.

RODENTICIDES AND BLOOD THINNERS

The haemorrhagic agent dicoumarol was announced to the world in 1940, and soon afterwards was being used in human volunteers to test for its efficacy as a treatment for thrombosis and acute heart failure. One of the other uses dicoumarol was tested for was as a rat killer, but in this respect it was found to be weak and unreliable. However, during the peak of research into the active compound in spoiled sweet clover, hundreds of 3-substituted 4-hydroxycoumarins had been synthesized and these were also studied for their effectiveness as rat poisons. Compound number 42 (3-phenyacetyl ethyl 4-hydroxycoumarin) was selected as the most potent, and when it was patented in 1948 it was named warfarin after the Wisconsin Alumni Research Foundation (WARF), which had sponsored the work.

Warfarin was initially marketed and used as a rat poison, the belief being that it was too poisonous to be used in humans. The unsuccessful attempted suicide of a naval recruit by ingesting a large quantity of warfarin and his subsequent full recovery, however, changed the minds of clinicians. Warfarin was found to be a far superior anticoagulant than dicoumarol, and when President Dwight Eisenhower suffered a heart attack in 1955 he was treated with the drug. So the cow killer was replaced by the rat poison, which soon became the standard treatment for long-term thrombotic conditions.

PAPAVERACEAE

The poppy family is one of the smaller families featured in this book. Even in its most recent circumscription, which includes the previously recognized fumitory family (Fumariaceae, now the subfamily Fumarioideae), it contains only around 800 species in roughly 40–45 genera. Although the morphology of plants belonging to either of the two former groups is distinct, they share several biosynthetic pathways for alkaloids. Most members of the poppy family are herbaceous, but there are a few that are shrubs or even small trees (such as *Bocconia* spp.). They are mainly found in north temperate regions, although there are also some species in western South America and South Africa.

BAKED GOODS FAIL DRUGS TEST

Humans use very few plants in the poppy family as food – their main uses are for medicine and as ornamentals. The most economically important species is the opium poppy (*Papaver somniferum*; see pages 200–201), which is cultivated for all three uses. Apart from being grown in gardens for its pretty flowers and structural capsules, the opium poppy contains several alkaloids in its latex (see box), most notably the analgesics codeine and morphine, the latter of which is processed to produce the illegal drug heroin. For this reason, the species must be among the plants that have caused most deaths in humans. Despite this record, its seeds are used in baking to add colour and crunch to bread and cakes. While they may contain only minute amounts of active alkaloids – not enough to cause any physiological effects – eating the seeds can possibly result in a positive drug test if large amounts are ingested.

BELOW **The flowers of bloodroot (*Sanguinaria canadensis*), a woodland plant, arise singly and are surrounded by a deeply lobed leaf that can reach up to 22 cm (9 in) across.**

BELOW RIGHT **Common ramping-fumitory (*Fumaria muralis*), a small European annual, has heads of bilaterally symmetrical, dark-tipped pink flowers.**

LEFT **The greater celandine (*Chelidonium majus*) is native to Europe and western Asia. This branching perennial is in a different family from lesser celandine (*Ficaria verna*; see page 125).**

THE ALKALOIDS

Members of the poppy family produce a large number of isoquinoline alkaloids derived from 1-benzyltetrahydro-isoquinoline precursors. This group of compounds have their largest diversity in this family. There are seven main types: morphinans, protoberberines, benzophenanthridines, protopines, cularines, phthalideisoquinolines and aporphines. The alkaloids are protective, being toxic to bacteria, viruses and fungi, as well as to insects. Larger herbivores are also deterred from eating the plants and so are rarely poisoned, and the plants can become serious weeds. A number of species are used in traditional herbal medicine, but care is generally advised.

Morphinans are restricted to the poppies (*Papaver* spp.), such as codeine and morphine that can be found in the opium poppy, and thebaine found in other species. Protoberberines are the most common type and include berberine. Benzophenanthridines, such as sanguinarine, are found in Mexican prickly poppy (*Argemone mexicana*; see pages 174–175) and have poisoned thousands of people through contamination of cooking oil. Unlike the morphinans in the opium poppy, these more toxic alkaloids are restricted to the seeds of the Mexican prickly poppy. The last two types of alkaloids are restricted to, or at least mainly found in, members of the subfamily Fumarioideae.

MEDICINAL LATEX

The sap is one of the characters that is used to differentiate between plants of the Fumarioideae subfamily and the Papaveroideae subfamily, which contains members of the more narrowly described Papaveraceae. In the former, the sap is watery, but in the latter it is a viscous yellow, orange, red or white latex. Opium is the dried white latex of the opium poppy, gathered by scoring the capsule. Brightly coloured latex is found in, for example, the greater celandine (*Chelidonium majus*), in which it is yellow to orange, turning red on exposure to air. This substance was traditionally used to treat warts, although this is now discouraged. The North American bloodroot (*Sanguinaria canadensis*) is named after its deep orange to red sap. Extracts of the rhizome are used in oral hygiene products for their antibacterial properties, but products containing the isolated sanguinarine have been withdrawn due to potential carcinogenic effects.

BELOW **The red latex from bloodroot (*Sanguinaria canadensis*) rhizomes was used by the Native Americans to dye baskets and fabric, and as face paint.**

Cooking oil contaminant – benzophenanthridine alkaloids

Some of the most serious poisonings occur when a toxic plant is eaten as a food. Usually in such cases the plant has been misidentified as an edible species and an individual or a family are affected by the mistake. But the Mexican prickly poppy contaminates whole crops, so the effects can reach entire communities.

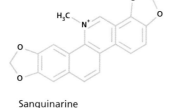

RIGHT **Sanguinarine is a benzophenanthridine alkaloid present in many members of the poppy family. After repeated ingestion, it may cause oedema and heart failure.**

Sanguinarine

A PRICKLY SITUATION

PLANT:
Argemone mexicana L.
COMMON NAMES:
Mexican prickly poppy, yellow thistle
FAMILY:
poppy (Papaveraceae)
TYPE OF TOXIN:
benzophenanthridine alkaloids (dihydrosanguinarine and sanguinarine)
SYMPTOMS OF POISONING IN HUMANS:
After a single ingestion:
DIGESTIVE: mild vomiting and diarrhoea

After repeated doses:
CIRCULATORY: hypotension (low blood pressure), tachycardia, congestive heart failure
NEUROLOGICAL: burning feet, tenderness of affected limbs
DIGESTIVE: vomiting, diarrhoea
SKIN: reddening of legs (erythema)
OTHER: swollen limbs, swollen areas that do not bounce back after the application of pressure (pitting oedema), low fever, breathlessness, sometimes glaucoma

The Mexican prickly poppy is a bluish-green annual herbaceous plant with prickly leaves, bright yellow sap and yellow flowers that have delicate petals. The plant protects itself not only by the presence of these prickles, but also by being unpalatable, so is rarely eaten by herbivores. It is native to central Mexico and southwards to Honduras, and is naturalized elsewhere as a pantropical weed, particularly in

LEFT **The Mexican prickly poppy (*Argemone mexicana*) can grow in inhospitable conditions and has become a successful, invasive weed in many regions that are sufficiently warm.**

areas of disturbed ground. The species is a particular hazard when it occurs among crops grown for their seeds, such as black mustard (*Brassica nigra*) in the cabbage family (Brassicaceae) (see pages 120–121), alfalfa in the legume family, and cereals such as wheat (*Triticum aestivum*) in the grass family. If not removed during harvesting, the Mexican prickly poppy seeds will cause the resulting food to be toxic.

The Mexican prickly poppy plant contains several alkaloids, including the benzophenanthridine dihydrosanguinarine and smaller amounts of the more toxic sanguinarine, in its seed oil (known as katkar or argemone oil). These two compounds cause dilatation of capillaries, increasing their permeability and enabling plasma to leak into the compartments around cells. Fluid builds up, causing the tissue to swell and forming oedema, with the lower limbs being the first affected. In the most serious cases, the proper function of the lungs and heart is compromised, with death usually the result of heart failure. In less serious cases, the swelling of the legs can remain for several months.

Epidemic dropsy

In India, where oil from black mustard seeds is widely used in cooking, contamination with Mexican prickly poppy oil has led to outbreaks of a condition known as epidemic dropsy (dropsy is a term once used for oedema; see page 198). The Mexican prickly poppy is an introduced species in India, where the first case of epidemic dropsy was reported in Kolkata in 1877. Outbreaks have occurred regularly since then, with one of the worst in 1998, when more than 3,000 people were admitted to hospital and more than 60 died. A smaller outbreak affecting three families was reported as recently as 2013.

Symptoms develop after contaminated oil has been eaten over several days, with the most severe effects seen in those who have consumed badly contaminated oil for longer periods. Contamination of the mustard oil is usually unintentional, although intentional contamination for financial gain is also possible. Mustard oil's pungent taste can mask the presence of the equally pungent katkar oil, and contamination levels as low as 1 per cent have caused epidemic dropsy. Improved control of bottled oil sold in India has seen a fall in the number of cases, but families involved in the cultivation of black mustard crops are still at risk.

ABOVE **The yellow-petalled flowers of Mexican prickly poppy (*Argemone mexicana*) are followed by prickly ellipsoid capsules that open to release numerous black seeds containing toxic alkaloids.**

Sieving for clues

Although animals avoid eating Mexican prickly poppy and other species in the genus *Argemone*, they have been poisoned when fed a diet that included the seeds. For example, in 1962 in South Africa, sheep died from eating contaminated wheat. Some people who ate the sheep that had died were themselves poisoned, suggesting that this may have been an example of secondary poisoning. It is more likely, however, that they were also eating the contaminated wheat. Records from the late 1940s show that it was common practice for some farmers to ask millers to set their sieves at a fineness that did not eliminate all weed seeds. Contaminated grain would then be fed to the farm animals and also ground into flour for the farm workers to eat. When the workers developed symptoms similar to epidemic dropsy, the link to Mexican prickly poppy was eventually made and active steps were taken to inform all concerned of the danger to both humans and animals.

CHAPTER 9

CELL POISONS

When we think of poisonous and toxic compounds, we usually think of those substances that cause acute drastic effects, and this is certainly the case if they act on individual cells to stop their ability to produce energy. Other compounds act more slowly, and some require repeated exposure, so there can be a delay before their deleterious effects are felt. Some can even increase the risk of developing cancer or have effects on foetal development, but as these contentious subjects would require a book of their own, this chapter only explores some examples of clearly high-risk compounds.

CELL POISONING: MECHANISMS OF ACTION

Some plants contain poisons that exert their potentially lethal activity by affecting fundamental processes present in every cell. When cellular energy production is destroyed, this can lead to the rapid onset of symptoms and death, or if the processes are slow, symptoms do not become obvious until extensive and possibly irreversible damage has been done. In some instances, it is very hard to connect plants and their poisons to fatal effects, such as in the case of carcinogenic plant compounds or those affecting foetal development.

LEFT **The fluoroacetic acid in prickly poison (*Gastrolobium spinosum*), an Australian shrub, has poisoned cattle.**

MESSING UP MOVEMENT

The large group of plants known as legumes contain many different toxins, several of which have already been discussed in this book. In other species, unusual amino acids that are not normally incorporated into proteins are present. These can have a variety of effects, with the disease neurolathyrism being among the most serious (see pages 186–187). In this case, the responsible amino acid mimics the common neurotransmitter glutamate in the human brain, effectively destroying brain cells involved in motor function and leading to movement disorders.

Among these aberrant amino acids is *beta*-methylamino-L-alanine (BMAA), present in cycads. It is thought to be involved in the development of a parkinsonism-like disease (see pages 188–189) by incorporation into human proteins, which cannot

TIME-DELAYED DELIVERY

We have already encountered the fact that the human body has a number of defensive mechanisms against being poisoned. So how can these defences be circumvented? One way is to produce compounds that need specific metabolic activation. Such substances will sneakily be absorbed into the body and produce lethal effects only after encountering enzymes that modify their structure. Among the most effective of such time-delayed poisons are those affecting cellular energy production, including fluoroacetic acid (see pages 180–181) and cyanogenic glycosides (see pages 182–185). In the latter, there is actually a 'double whammy' activation, whereby metabolic processes happen both in the plant when it is grazed, and if that is not enough to deter a herbivore, further activation continues inside the animal's body, poisoning it as cyanogenic precursors are metabolized, and possibly leading to the disease konzo (see page 185).

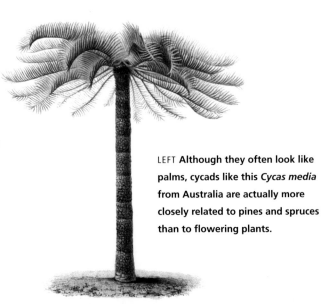

LEFT **Although they often look like palms, cycads like this *Cycas media* from Australia are actually more closely related to pines and spruces than to flowering plants.**

RIGHT **Cycads have different female and male plants, easily distinguished by the cones they bear. Here are male cones of *Cycas media*.**

then function normally, leading to a staggered gait and, sometimes, dementia. Cycads also contain a compound similar to the cyanogenic glycosides, but instead of releasing poisonous cyanide it produces formaldehyde and a methylating substance. By themselves, and by depleting the body of protective glutathione, these cause liver damage. It has been proposed that such glutathione depletion is also responsible for the development of the now rare disease milk sickness, caused by drinking the tainted milk of cows grazed on tremetol-containing plants (see box, page 191).

SLOW BUT DEADLY

Many organisms rely on glycosylation of proteins (the attachment of different sugars) to ensure the proper function of enzymes, and also to create labels or 'address tags' that then allow the transportation of proteins to the correct location in the cell. Several plants, particularly the locoweeds (see pages 190–191), contain alkaloids that mimic sugars and will inhibit the enzymes responsible for this addition and labelling. This can lead to accumulation of sugars and formation of unnecessary storage vacuoles. The effect of this is twofold: enlargement of the affected cells, which then encroach on the neighbouring cells; and a shortage of the correctly labelled proteins. Both paths will lead to the diminishing function of other cells and can, for example, affect the brain (causing motor disease) and the heart (causing congestive heart failure). Other slow-acting poisons are those that increase the risk of developing cancer by indiscriminately binding to deoxyribonucleic acid (DNA) and thus potentially causing fatal mutations. In this chapter, this is exemplified by bracken (*Pteridium aquilinum*; see pages 192–193), which also poses an additional risk when consumed raw as it contains an enzyme that breaks down thiamine, causing acute vitamin B_1 deficiency. Deleterious effects on DNA are especially problematic during foetal development and can cause lethal malformations (see box, page 193).

BELOW **The beautiful Sturt's desert pea (*Swainsona formosa*) from southern Australia belongs to the genus that gave its name to swainsonine, an alkaloid that causes locoism in livestock.**

The poison leaf – fluoroacetic acid

During the late 1830s, a pioneer movement known as Voortrekkers, comprising Dutch descendants in South Africa's Cape Colony, travelled inland in search of a better life. On entering the area that would become known as Transvaal, they encountered a plant that was especially poisonous to their cattle and sheep. They called it *gifblaar*, which translates from Afrikaans as 'poison leaf'.

Injecting chemical pain

PLANT:
Dichapetalum cymosum (Hook.) Engl.
COMMON NAMES:
gifblaar, poison leaf
FAMILY:
gifblaar (Dichapetalaceae)
TYPE OF TOXIN:
fluoroacetic acid

SYMPTOMS OF POISONING IN HUMANS:
DIGESTIVE: abdominal pain, vomiting, diarrhoea
NEUROLOGICAL: sweating, confusion, agitation, coma
CIRCULATORY: arrhythmia, hypotension
OTHER: uncoordinated muscle movements, paralysis

Fluoroacetic acid   Fluorocitric acid

ABOVE **Fluoroacetic acid present in plants will, after absorption into the body, be converted into fluorocitric acid, a compound that kills by halting the energy production of mitochondria.**

Gifblaar is a peculiar-looking plant, growing in clusters of low shoots with a few leaves and flowers close to the ground. It has been described as an 'underground tree', as clusters of what seem to be a number of plants are often a single individual connected underground through branched roots. This immense root system taps into water deposits far below the surface, enabling the plant to produce leaves even before the summer rainfall. Such spring foliage was previously considered to be especially poisonous to cattle, although the greater incidence of spring poisonings may, in fact, be due to the lack of other suitable grazing at that time of year.

It was originally thought that the poison in gifblaar was cyanide, due to the similar presentation of symptoms (a rapid onset of diminishing consciousness and staggering movements), but that particular toxin has not been detected in the plant. The actual toxic compound, fluoroacetic acid, was identified in 1944, more than a century after the start of the Voortrekkers' migration. This compound, in the form of its sodium salt, had earlier, in 1942, been independently synthesized and proposed as a promising new rat poison under the name 'compound 1080'. The story comes full circle, as the gifblaar relative called ratsbane (*Dichapetalum toxicarium*, syn. *Chailletia toxicaria*) had extensive use as a rodenticide in western Africa around

LEFT **These shoots may be a single gifblaar (*Dichapetalum cymosum*) plant, connected by a root system. Early settlers of the Transvaal found that the leaves were poisonous to livestock.**

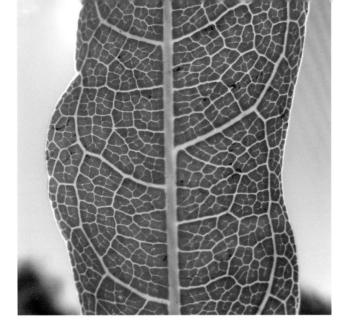

ABOVE A gifblaar (*Dichapetalum cymosum*) leaf, showing its uneven margin and network of lighter veins, which transport water to the cells and glucose away to the rest of the plant.

the turn of the nineteenth century (see box). Due to its high toxicity, compound 1080 has been used not only to combat rat infestations, but also, more controversially, in efforts to control non-native animals in Australia and New Zealand, and protect domesticated animals in various parts of the world.

Parallel evolution

The gifblaar plant is not the only plant that contains fluoroacetic acid. The compound has been reported from more than 50 different species in a handful of genera and families in a case of parallel evolution. In Brazil, cattle succumb to poisoning by the widespread erva-de-rato (*Palicourea marcgravii*) from the coffee family (Rubiaceae), and the substance has been detected in several relatives of this species. And in southwestern Western Australia, the genus *Gastrolobium*, commonly called poison pea and in the legume family (Fabaceae), was responsible for large losses of cattle in the mid-nineteenth century. The economic impact was deemed so serious that eradication programmes were instigated in many regions, now leaving several of the 100-plus species of *Gastrolobium* rare or threatened by extinction.

Stopping the cycle

Fluoroacetic acid is not intrinsically poisonous, but its obvious similarity to acetic acid means that some enzymes in animals (including humans) do not distinguish between the two compounds. In animal cells, energy in the form of adenosine triphosphate (ATP) is produced by the mitochondria in the series of reactions called the citric acid cycle, and by being incorporated into fluorocitric acid a potent inhibitor of this chemical pathway is produced. If the ingested amount affects enough mitochondria, the body's stored energy will soon be depleted, leading to rapid and fatal poisoning.

Mystery of the 'broken back'

In 1903, a 24-year-old male labourer was admitted to the Colonial Hospital in Freetown, Sierra Leone, after severe vomiting and trembling, which rapidly developed into a complete inability to walk and difficulty in using his arms. With the help of a translator, doctors concluded that he had ingested bait aimed to kill rats in the form of fish sprinkled with the powdered seeds of ratsbane. The young man was paralyzed for a fortnight, but recovered slowly and was discharged from hospital after two months. The medical officer at the hospital drew parallels between this case and a mysterious and often fatal illness haunting the area, which affected mainly younger adults and started with lower limb paralysis. Traditionally, this 'broken back' disease was said to be caused by the devil or witchcraft, but ratsbane poisoning seems to be a more rational explanation.

BELOW The erva-de-rato (*Palicourea marcgravii*) shrub has heads of small, vividly coloured flowers. It is responsible for a significant number of cattle deaths in Brazil each year, due to the presence of fluoroacetic acid and other toxins.

Death by cyanide – cyanogenic glycosides

Cyanogenic glycosides (of which there are around 300 different types) are widespread in plants, including around 3,000 species from more than 100 families, many of them edible. Single ingestions can cause poisoning that will usually resolve with no lasting consequences, but may be fatal if a large quantity is eaten. However, even smaller amounts eaten over an extended period, usually as part of the diet, cause chronic health conditions, including the paralyzing konzo.

Releasing cyanide

Cyanogenic glycosides are stored in the cell vacuoles of the plant and have very low toxicity. If the cells are damaged, however, such as through chewing, the compounds are exposed to the enzyme *beta*-glucosidase. This enzyme quickly splits the sugar from the cyanogenic glycoside and the resulting nitrile then splits further, either spontaneously or facilitated by a second enzyme, nitrilase, into an aldehyde and hydrogen cyanide (HCN or hydrocyanic acid, commonly referred to as 'cyanide'). Once the plant matter is swallowed, further cyanide from initially intact plant cells can be released in the stomach due to the action of *beta*-glucosidases from the microbial gut flora. The toxic cyanide is readily absorbed into the circulatory system and transported around the body, where it interferes with cellular respiration, leading to cell death.

BELOW **Apricots (*Prunus armeniaca*), probably originating in central Asia, are now grown worldwide as an important fruit crop. The sweet orange flesh encloses a stone with a poisonous kernel.**

Bitter kernels

PLANT:
Prunus armeniaca L.
COMMON NAMES:
apricot
FAMILY:
rose (Rosaceae)
TYPE OF TOXIN:
cyanogenic glycosides
(prunasin, amygdalin)

SYMPTOMS OF POISONING IN HUMANS:
CIRCULATORY: palpitations, hypotension, cardiovascular collapse
NEUROLOGICAL: anxiety, headache, dizziness, confusion, decreased consciousness, seizures, coma
DIGESTIVE: nausea, vomiting, abdominal cramp
OTHER: weakness, paralysis, respiratory failure

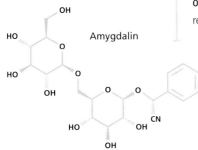

LEFT **Amygdalin is a major cyanogenic glycoside in the rose family, especially in bitter almonds and apricot kernels.**

Many members of the rose family (Rosaceae) are important sources of edible fruit, despite the deadly secret that they hide in their seeds or kernels. Cyanide can be released from the seeds of apples (*Malus domestica*) and the kernels of cherries (*Prunus avium*), plums (*P. domestica*), peaches (*P. persica*) and almonds (*P. amygdalus*), of which there are both bitter and sweet varieties. However, it is apricot (*P. armeniaca*) kernels that have caused the most poisonings, since their promotion as a 'cancer cure'.

The apricot is a small tree that originates in Kyrgyzstan and China's Xinjiang province. It was domesticated approximately 5,000 years ago and was introduced to the Caucasus and Iran, from where it spread into Europe, either during the period of Alexander the Great's campaigns (334–323 BC) or via the Silk Road. The Spanish took the species to North America, and it is now cultivated in warm regions around the world. Apricot fruit are drupes, having a fleshy exterior and a single stone that encloses the seed kernel. The sweet flesh of the fruit is edible, being eaten fresh or dried, but the bitter kernel contains variable amounts of cyanogenic glycosides, particularly amygdalin (*D*-mandelonitrile-*beta*-D-glucoside-6-*beta*-glucoside).

RIGHT **Each apricot fruit contains a single stone that can be split open to reveal the toxic kernel within.**

Vitamin B$_{17}$

Apricot kernels occasionally cause poisoning, sometimes fatal, in areas of the world where the fruit is grown as a crop. This is usually as a result of accidental ingestion by children, but also when they have been used to flavour food without proper processing. However, more recently another cause of poisoning from apricot kernels has arisen. Amygdalin, particularly in a chemically treated form called laetrile (also misleadingly known as vitamin B$_{17}$), was proposed as a treatment for cancer in the 1950s and by 1978 had been used by 70,000 patients in the United States alone. A lack of clinical evidence to support its use, however, as well as its potential toxicity, led to its ban in that country and also in Europe, Canada and elsewhere. Despite this, apricot kernels themselves have become a popular alternative therapy and health food. Being a natural product, these kernels contain varying concentrations of amygdalin, which in some instances reaches very high levels, and as people eat large numbers of the kernels at a time, toxic amounts can easily be ingested.

Prussic acid poisoning

Prussic acid poisoning gets its name from an older term used for hydrogen cyanide. It develops quickly in livestock that eat plants containing cyanogenic glycosides, and can affect entire herds. Ruminant animals, such as cows, are more susceptible to poisoning due to the microbial fermentation in the rumen. Some grasses (Poaceae), particularly sorghum or great millet (*Sorghum bicolor*, syn. *S. vulgare*, see photo below), can cause this type of poisoning. Sorghum is native to central Africa and is now grown widely, particularly in semi-arid regions of the world, for its grain, which is a staple food for more than 500 million people. The leaves and stems of sorghum contain the cyanogenic glycoside dhurrin, so it is important to keep animals away from the plants, particularly when concentrations of the toxin are at their highest: in young plants, following frost or during periods of drought.

The pros and cons of cassava

PLANT:
Manihot esculenta Crantz
COMMON NAMES:
cassava, manioc, tapioca
FAMILY:
spurge (Euphorbiaceae)
TYPE OF TOXIN:
cyanogenic glycosides (linamarin, lotaustralin)

SYMPTOMS OF POISONING IN HUMANS:
After a single ingestion:
as for *Prunus armeniaca* (see page 183)
After repeated doses:
NEUROLOGICAL: paralysis of lower limbs, damage to eyesight, deafness
OTHER: weakness

ABOVE **The cyanogenic glycoside linamarin is present in many plants used for food, such as the cassava root.**

The cassava (*Manihot esculenta*) plant is just one of the toxic members of the Euphorbiaceae family (see pages 114–115). This perennial woody shrub is native to South America but was introduced to Africa by the Portuguese in the seventeenth century. Despite being poisonous, its large, elongated tubers now form the staple diet of people living in many subtropical and tropical countries.

There are several advantages to growing cassava, not least that it requires considerably less human investment per calorie than many other crops, and can tolerate periods of drought or flooding when other crops would fail. The powerful combined deterrent of a bitter taste and the sudden release of cyanide means that, like other plants containing both cyanogenic glycosides and the enzymes that release cyanide, it is very resilient to pests. But the toxic compounds in this remarkable crop do have a considerable downside: they can cause great harm if the tubers are not processed properly before being eaten.

Sweet and bitter

The varieties of cassava can be broadly divided into sweet and bitter. While the type of cyanogenic glycoside that they contain is the same – mainly linamarin and a small percentage of lotaustralin – the quantities in the tubers can vary enormously, from 15–100 mg cyanide equivalent per kilogram in sweet varieties to 500 mg in bitter tubers, and can increase further in response to drought. Sweet cassava is normally used as a food crop, but bitter cassava is sometimes grown in preference despite the increased risk of poisoning because of its greater resilience.

When processing cassava, peeled whole tubers are normally soaked in water to reduce the cyanide content to safe levels. This may be effective for the relatively low levels of cyanogenic glycosides found in sweet varieties grown under normal conditions, but it is not sufficient for bitter varieties. For these, processing involves grinding peeled tubers into flour, which ensures the enzymes and cyanogenic glycosides can mix and release cyanide, followed by soaking the flour in water to ensure hydrogen cyanide evaporates away safely.

Acute and chronic

Eating a single meal of insufficiently processed cassava will cause acute cyanide poisoning, but as cases are mostly non-fatal and frequently occur in geographically remote areas where they are not unusual, such incidents are under-reported. Poisoning is usually the result of eating cassava tubers, particularly bitter varieties, that have been insufficiently prepared. Cases were unknown in Venezuela until 1992, when several children were severely poisoned. More recently,

RIGHT **Plantation of young cassava (*Manihot esculenta*) plants showing the palmate (hand-shaped) leaves. The plants can eventually reach 1–3 m (3–10 ft) in height.**

LEFT **Each cassava (*Manihot esculenta*) plant can produce a large cluster of tuberous roots, providing a significant crop for the farmer, but its valuable carbohydrate is protected by cyanogenic glycosides.**

BELOW **The preparation of cassava, exemplified by the Maroon people of Suriname, is a multistep process of pounding and washing that leaches the cyanide away and ensures the tubers are safe to eat.**

dramatic increases in food prices drove people to cheaper, unfamiliar sources of nutrition, including bitter 'industrial' cassava, resulting in 11 deaths in a few months. In areas of Africa, a chronic neurological condition known as konzo, which is similar to neurolathyrism from species of grass pea (*Lathyrus* spp.; see pages 186–187), also results from eating cassava (see box).

Konzo

Konzo (cassava-associated spastic paraparesis) is an irreversible medical condition thought to be caused by eating cyanogenic glycosides from improperly processed cassava in a regular diet that is low in protein. Konzo means 'tied legs' in the language of the Yaka people of southwestern Democratic Republic of the Congo and is a very good description of the symptoms, which can leave victims unable to walk. Symptoms come on quickly, and individuals can wake in the morning to find their lower limbs paralyzed and inflexible.

Outbreaks of konzo occur in eastern, central and southern Africa, usually during times of drought. The specific cause of the condition is still not clear, but there may be less sulfur in the body to aid cyanide detoxification into thiocyanate (see box, page 35) when the diet is low in protein. Other theories are that nerve damage results from high concentrations of thiocyanate produced after eating bitter cassava, or that the actual cyanogenic glycosides themselves cause konzo without releasing any cyanide at all.

Neurolathyrism – non-proteinergic amino acids

Many members of the legume family (see pages 102–103) provide important sources of dietary protein, but in some cases precautions have to be taken to make them safe to eat. For example, kidney beans (*Phaseolus vulgaris*; see pages 146–147) contain proteins known as lectins and need to be cooked thoroughly before consumption. Here, we look at the grass pea (*Lathyrus sativus*), in which a free amino acid causes toxicity.

Not so sweet pea

PLANT:
Lathyrus sativus L.
COMMON NAMES:
grass pea, chickling vetch
FAMILY:
legume (Fabaceae)
TYPE OF TOXIN:
non-proteinergic amino acids (*beta-N*-oxalylamino-L-alanine (BOAA), oxalyldiaminopropionic acid (ODAP))

SYMPTOMS OF POISONING IN HUMANS:
NEUROLOGICAL: spastic gait, progressive difficulty in walking, contracture paralysis
DIGESTIVE: diarrhoea, vomiting
OTHER: respiratory arrest

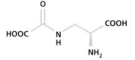

beta-*N*-oxalylamino-L-alanine

ABOVE **The amino acid *beta-N*-oxalylamino-L-alanine is responsible for a paralysing disease called neurolathyrism.**

Grass pea is a branching annual herbaceous plant that can reach 170 cm (5.5 ft) in height and climbs by means of tendrils at the end of each leaf. The classic pea-shaped flowers are blue, red, pink or white, and are followed by flattened pods measuring up to 5.5 cm by 2 cm (2 in by ³⁄4 in). Each pod contains up to seven wedge-shaped seeds, 4–7 mm (³⁄16–¹⁄4 in) in diameter. The grass pea's place of origin is uncertain, but there is archaeological evidence that it was domesticated in the Balkan region around 6000 BC. Its cultivation has now spread throughout Asia, southern Europe, north Africa and elsewhere, due to its tolerance of extreme environments, ease of cultivation and ability to grow in poor soil through nitrogen fixation in root nodules.

LEFT **Grass pea (*Lathyrus sativus*) seeds can be eaten if they are only a small part of the diet and are prepared properly, but they have caused poisoning during drought conditions.**

FAMINE FOOD

Grass pea is usually consumed as just a small part of the diet, but during periods of drought it is often the only crop to survive and can of necessity become a significant source of calories. At such times, normal methods of preparation that detoxify the grass pea, such as soaking and boiling in water, and mixing with wheat flour, are no longer a possibility. If this continues for several weeks or months, a neurological condition called neurolathyrism can develop in a small proportion of people. Symmetrical degeneration of neurons in the spinal cord and cells in the area of the cortex that controls the legs results in spastic paraplegia. The progression is very slow and can be halted by improving the diet, but unfortunately it is incurable. These symptoms resemble those of konzo, caused by chronic ingestion of cassava (see box, page 185), while a slightly different form of paralysis results from eating coyotillo (*Karwinskia humboldtiana*; see box).

Although the debilitating but non-fatal condition of neurolathyrism was once more widespread, it is now generally seen only in parts of Asia and Africa. During severe famines in Ethiopia in 1976–1977, at least 2,600 people out of a population of just over a million developed neurolathyrism. And following the drought of 1995–1996, a neurolathyrism epidemic started in February 1997 and some 2,000 people were known to have developed the condition within a year. Approximately 2 per cent of those who consume a diet rich in grass pea develop the condition, with males, adolescents and children being most at risk. The incidence could be dramatically reduced by providing aid in the form of cereals to add to the grass pea and water for detoxification, in addition to drinking water.

RIGHT **The grass pea (*Lathyrus sativus*, illustrated here) resembles the sweet pea (*L. odoratus*), which is grown for its sweet-smelling flowers; sweet pea seeds contain a different toxin and long-term feeding has poisoned turkeys.**

Coyotillo curse

In 1918, the poisoning of 106 Mexican soldiers, 20 per cent of whom died, was the first report of a large human intoxication by coyotillo, a shrub in the buckthorn family (Rhamnaceae) that grows in semi-desert habitats in north and central Mexico and Texas in the United States. From November to February, coyotillo is covered in dark purple fruit that are sweet to eat. The seeds contain neurotoxic anthrones such as peroxisomicines, which damage the peripheral nerves through demyelization.

Other large coyotillo poisonings have occurred subsequently, and also individual poisonings – at least several each year in the 1980s, almost exclusively in children. If a large amount is eaten, diarrhoea and vomiting are initially the only symptoms, but after a delay of anything between one day or four weeks, a flaccid, symmetrical, ascending paralysis of the limbs can develop, which may end in death by respiratory arrest. The length of the delay and the severity of the resulting symptoms depend on the amount that is eaten, and whether the fruit have been eaten only once or repeatedly over a number of days. Those who survive usually make a full but slow recovery.

TOP **Branch of a coyotillo (*Karwinskia humboldtiana*) shrub growing in Texas. Plants can reach 4–6 m (13–20 ft) in height, and have opposite pairs of leaves and one to three small flowers growing from the leaf axils.**

Cycads and cyanobacteria

The cycads are a group of slow-growing tropical and subtropical palm-like trees that have barely changed since before the time of the dinosaurs – cycad fossils date back to the Late Palaeozoic era, 290–265 million years ago. Their resistance to hurricanes and droughts is part of the reason for their continued survival to the present day. Over the centuries, humans have used cycads for food and medicine, but the toxins they contain mean they have to be processed before they are consumed. Even then, there can be long-term consequences.

POISON FROM BEFORE THE DINOSAURS

PLANT:
Cycas revoluta Thunb.
COMMON NAMES:
sago cycad
FAMILY:
cycad (Cycadaceae)
TYPE OF TOXIN:
azoxymethanol glycosides (cycasin, neocycasins); non-proteinergic amino acids (*beta*-methylamino-L-alanine (BMAA))

SYMPTOMS OF POISONING IN HUMANS:
After a single ingestion:
CIRCULATORY: tachycardia
NEUROLOGICAL: headache, dizziness, weakness
DIGESTIVE: abdominal pain, severe vomiting, diarrhoea
OTHER: liver damage
After repeated doses:
NEUROLOGICAL: progressive paralysis, Parkinson-like symptoms, sometimes dementia

beta-methylamino-L-alanine

Cycasin

ABOVE **All cycads contain a compound similar to cyanogenic glycosides, cycasin, and some may contain *beta*-methylamino-L-alanine, produced by cyanobacteria.**

EMERGENCY RESERVES

The sago cycad (*Cycas revoluta*), a member of one of the two families of cycad (Cycadaceae, the other one being Zamiaceae), is often called the 'sago palm' but should not be confused with the true sago palm (*Metroxylon sagu*) in the palm family (Arecaceae). Native to Japan, it is probably the most widely cultivated cycad. Various parts of this and other cycads are eaten by humans, usually when other crops have been destroyed by natural disasters or as a stop-gap during seasonal shortages, but also as a staple part of the traditional diet in many regions. The young leaves may be eaten as a vegetable, but it is the seeds and also the stem pith that are most often used as they, after a long detoxification process, provide a flour with a high starch content.

LEFT **Crown of the sago cycad (*Cycas revoluta*), with a head of developing seeds attached to small leaf-like structures, and surrounded by rigid palm-like leaves up to 1.5 m (5 ft) long.**

On the breadline

During the Boer War (1899–1902), a group of Boers commanded by Jan Christiaan Smuts (1870–1950) found themselves extremely short of food in the Zuurberg Mountains and decided to eat the seeds from 'a strange growth known as "Hottentot's bread" (*Encephelartos Altensteinii* [sic]), a large fruit not unlike a pineapple'. Botanists have identified this as a different species, the breadpalm (*Encephalartos longifolius*). One of the soldiers tried it and liked the flavour, and soon many others in the company followed suit. It resulted in severe poisoning, with more than half the men 'groaning and retching on the ground in agony'. General Smuts was worse than the rest, lying comatose for a period. Fortunately, no one died and those who were well were able to fight off the enemy while their comrades recovered over the following days.

RIGHT **Breadpalm (*Encephalartos longifolius*) is a rare cycad of Eastern Cape Province in South Africa. It produces large female cones with red nuts covered by yellowish scales.**

When Europeans first encountered cycads during their voyages of discovery, they were unaware of their toxicity. During Captain James Cook's first voyage to Australia in 1770, the botanist Joseph Banks noticed that several crew members became violently ill after eating nuts from *Cycas media*, and General Jan Smuts and his troops fell foul of the breadpalm (*Encephalartos longifolius*, Zamiaceae) during the Boer War (see box). In Honduras, it has been documented that the roots of camotillo (*Zamia furfuracea*, Zamiaceae) were used in unlawful poisonings. Improper processing of cycad plants before consumption, either as a food or traditional remedy, leaves the azoxymethanol glycosides they contain at toxic levels and is now the usual cause of acute poisoning. A second toxin, *beta*-methylamino-L-alanine (BMAA), which is particularly concentrated in the seeds and root nodules, is not removed by the processing, but only takes effect if the plant is eaten on repeated occasions.

RIGHT **Eating improperly processed seeds of the sago cycad (*Cycas revoluta*) and other cycad species can cause acute poisoning, including vomiting, headache and dizziness.**

GUAM PARKINSONISM

The Chamorro people of Guam, in the Mariana Islands of the Pacific, have a high incidence of a neurological disease called amyotrophic lateral sclerosis/parkinsonism–dementia complex (ALS/PDC). This has been linked to their consumption of fadang (*Cycas micronesica*) seeds, from which they produce a flour that contains the toxin BMAA. This compound has also been found in another component of the traditional Guam diet, fanihi, or flying foxes (*Pteropus mariannus*), which consume the fadang seeds and concentrate the BMAA in their body fat. The flying foxes are now close to extinction due to overhunting and the use of fading-seed flour has also declined with the introduction of Western food; correspondingly, the incidence of ALS/PDC among the human population has also decreased. There is another twist to this story. Fadang does not in fact produce the BMAA; this is done by symbiotic nitrogen-fixing cyanobacteria (such as *Nostoc* spp.) in the root nodules of the cycad, and is then accumulated in the seeds.

Locoweeds and milk sickness – swainsonine

Cattle, sheep, horses and other large animals graze on plants and so ingest large quantities, particularly of leaves, making them susceptible to poisoning by species not usually eaten by humans. The range of plants that can poison livestock is quite extensive. Some have been mentioned elsewhere in this book, such as the larkspurs (*Delphinium* spp.; see pages 48–49), which can kill after a single meal, and others that are poisonous if eaten over an extended period, such as the pyrrolizidine alkaloid-containing common ragwort (*Jacobaea vulgaris*) and comfrey (*Symphytum* spp.) (see pages 164–167). Here, some other plants that poison livestock following extended consumption are explored, including an unusual route of human exposure that took some time to unravel.

ABOVE **The white locoweed (*Oxytropis sericea*) grows in hills and mountains in the southwest United States. Swainsonine levels vary between plants and the toxin may even be completely absent.**

A few, mainly North American, species of milkvetch (*Astragalus* spp.) and oxytrope (*Oxytropis* spp.), two large genera in the legume family, are known as 'locoweeds'. Unlike the majority of species of milkvetch and oxytrope, the locoweeds contain the toxic indolizidine alkaloid swainsonine, and pose a serious threat to animals grazing on the rangelands where they grow. Although milkvetch and oxytrope species are used as a coffee substitute and in herbal remedies, they have not been associated with any harmful effects in humans.

Locoweed disease

PLANT:
Astragalus spp. and *Oxytropis* spp.
COMMON NAMES:
locoweeds
FAMILY:
legume (Fabaceae)
TYPE OF TOXIN:
indolizidine alkaloids (swainsonine and swainsonine *N*-oxide)

SYMPTOMS OF POISONING IN ANIMALS AFTER REPEATED DOSES:
NEUROLOGICAL: depression, ataxia (lack of voluntary coordination of muscle movement)
DIGESTIVE: weight loss
SKIN: rough coat
OTHER: abnormal gait and posture, muscle weakness (spastic paresis) in hind limbs, difficulty in standing

Swainsonine

LEFT **Swainsonine is an indolizidine alkaloid found in many legumes, and is probably produced by an endophytic fungus.**

Cattle, goats, horses and other animals that consume locoweeds for an extended period can develop locoweed disease, or locoism. In Australia, the plants in the related Darling pea genus (*Swainsona* spp.) that contain swainsonine cause a similar disease known as pea struck. Locoism, a chronic metabolic disease, is characterized by various neurological and behavioural disorders, as well as infertility, stillbirth and birth of weak offspring. Swainsonine can chemically imitate the structurally similar sugar mannose, affecting the enzymes that attach this sugar to other proteins to ensure their proper function and localization in the cell. This leads to accumulation of mismatched proteins, forming vacuoles that enlarge the cell and compress surrounding cells, further impairing their function.

The swainsonine is probably mainly produced by fungal endophytes (*Alternaria* sect. *Undifilum*) that reside in the locoweeds. The endophyte is found throughout the plant, but in seeds it is restricted to the seed coat and probably infects the seedling during germination. Swainsonine also occurs in some other plants that have poisoned livestock in Brazil and Australia, including *Sida carpinifolia* (probably *S. rhombifolia*) in the mallow family (Malvaceae) and members of the morning glory family (Convolvulaceae), including several species of morning glory (*Ipomoea* spp.) and moita de calango (*Turbina cordata*). Interestingly, some other members of the morning glory family are known to be poisonous due to the presence of endophytic fungi that produce ergot alkaloids (see pages 88–89).

ABOVE **The purple or Lambert's locoweed (*Oxytropis lambertii*) grows on the short- and mid-grass prairies of the south and western United States.**

Milk sickness

When Europeans started to settle in the Midwest region of the United States in the 1800s, they and their livestock began to fall ill. The animals developed violent trembling when they were forced to move or became agitated, and the disease became known as trembles. People who drank the milk of affected animals developed so-called milk sickness, and it is estimated that in some areas of Indiana and Ohio 25–50 per cent of the deaths of early settlers were caused by this condition. One casualty in 1818 was Nancy Hanks Lincoln, whose son, nine years old at the time, would become President Abraham Lincoln.

It took some time to identify white snakeroot (*Ageratina altissima*, syn. *Eupatorium rugosum*) as the cause of trembles. Although the plant was initially suggested as the culprit in the 1830s, this was only confirmed in the early 1900s. This member of the daisy family (Asteraceae) grows in moist, shaded areas, such as along stream beds and near tree lines. Animals do not show any signs of being poisoned

ABOVE **Nowadays, human poisoning by white snakeroot (*Ageratina altissima*) is rare due to industrial milk production, but it is an historically interesting killer plant.**

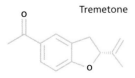

LEFT **Tremetone and similar compounds are responsible for the toxicity of white snakeroot (*Ageratina altissima*).**

until they have been eating white snakeroot for one to three weeks, and symptoms finally progress to chronic degeneration of the skeletal muscles. Benzofuran ketones, including tremetone, are at least partly responsible for the toxicity of white snakeroot, and they are also found in another member of the daisy family, the rayless goldenrod (*Isocoma pluriflora*, syn. *Haplopappus heterophyllus*), which causes a similar disease in grazing animals.

Causing cancer – ptaquiloside

Some of the defensive compounds produced by plants can be carcinogenic, causing cancer, or even teratogenic, causing birth defects in offspring if the mother eats them while she is pregnant (see box). Thankfully, there are relatively few plants whose toxins take such a long view, as fast-acting poisons are a more effective way to deter herbivores.

By hook or by crook

PLANT:
Pteridium aquilinum (L.) Kuhn
COMMON NAMES:
bracken, bracken fern
FAMILY:
bracken fern (Dennstaedtiaceae)
TYPE OF TOXIN:
terpene glycosides (ptaquiloside), enzyme (thiaminase), cyanogenic glycosides
SYMPTOMS OF POISONING IN ANIMALS:
After a single ingestion: in cattle and sheep, widespread haemorrhages and susceptibility to infection (as a result of bone marrow suppression)
After repeated doses: in horses, thiamine (vitamin B1) deficiency, leading to drowsiness, laboured breathing, unsteady gait, tremors, staggers, arching back, eventual recumbency with convulsions
CARCINOGENESIS: in cattle, bovine enzootic haematuria, leading to blood in the urine (haematuria) and bladder tumours – in cattle and sheep, upper digestive tract carcinomas

Ptaquiloside

LEFT **The terpene glycoside ptaquiloside may be carcinogenic, thus making bracken (*Pteridium aquilinum*) a potential but slow killer plant.**

BELOW **Bracken (*Pteridium aquilinum*), sometimes called eagle fern, is a nearly cosmopolitan plant growing in temperate and subtropical regions, where it poses a risk to grazing animals.**

Bracken is a fern, reproducing by spores, and with a spreading rhizome and branched fronds reaching 1.5 m (5 ft) in height. When treated as a single variable species, it is one of the most abundant plants on Earth. Its rhizome enables it to invade disturbed areas of land and dominate other vegetation. Bracken produces a cocktail of compounds that can provide an effective defence against attack by insects and larger herbivores. Unfortunately, however, the young fronds (known as croziers due to their similarity in shape to a bishop's crook) are still eaten by grazing animals, even though they contain some of the most toxic plant compounds.

THIAMINASE AND PTAQUILOSIDE

Bracken causes a number of different syndromes in animals. Horses that have eaten bracken for one or two months can show signs of thiamine (vitamin B_1) deficiency, such as weakness, tremors and incoordination. This is due to the presence of the enzyme thiaminase, which inactivates thiamine that is necessary for the correct functioning of the peripheral and central nervous systems.

Cattle and sheep can develop cancer of the upper digestive tract, and bleeding associated with decreased production of blood cells by the bone marrow (bone marrow suppression), while cattle are also susceptible to tumours of the bladder. The compound responsible for these activities is a terpene glycoside, ptaquiloside, which is converted in the liver to a carcinogenic metabolite. In sheep, ptaquiloside is also responsible for a condition known as bright blindness, which causes so-called star-gazing behaviour.

HUMAN HAZARDS

Humans are thought to have eaten bracken since prehistoric times, especially the rhizomes in winter and the croziers in spring. The rhizomes were eaten in Scotland during the First World War, and the croziers are still eaten today in Japan, China, parts of South America, Canada and elsewhere. The young fronds are boiled in water treated with soda ash (sodium carbonate), which reduces but does not eliminate their toxicity. An association has been identified between consumption of bracken croziers and development of cancer of the upper gastrointestinal tract in Japan and Brazil.

Eating meat or milk from cows that have grazed on bracken has also been suggested as another potential route of exposure to bracken toxins for humans. Ptaquiloside has been detected in milk from cows fed bracken or grazed on bracken-dominated vegetation, and people living in such areas (for example, in Costa Rica) have been found to have a higher rate of digestive tract cancer. Consuming milk directly from the farm is thought to be an important route of exposure for remote rural communities. Industrial milk production, in which milk pooled from different sources is pasteurized, reduces the risk to negligible levels.

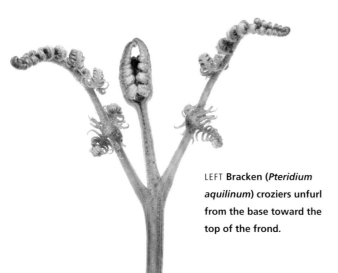

LEFT **Bracken (*Pteridium aquilinum*)** croziers unfurl from the base toward the top of the frond.

The not-so-mythical cyclops

In Greek mythology, the cyclops were giant-like monsters with a single eye in the middle of their forehead who were said to labour in the workshop of the god Hephaestus in the heart of the Mt Etna volcano. This has inspired the term cyclopia, used for animals born with a birth defect characterized by the absence of structures in the midline of the face, the most noticeable feature being a single eye. The cause is usually genetic, but occasionally there is an external factor.

In the western United States, cyclopia was a frequent occurrence in sheep flocks grazed on subalpine meadows, but went unreported until an epidemic in the 1950s. The cause was traced to the California false hellebore (*Veratrum californicum*) in the trillium family (Melanthiaceae). Studies showed that some populations of the plant contained the teratogenic compound cyclopamine (11-deoxojervine), which promoted cyclopia if eaten by pregnant sheep between the 10th and 15th days of gestation.

TOP The California false hellebore (*Veratrum californicum*) and related species in the trillium family growing in the northern hemisphere contain alkaloids that may cause cyclopism.

CHAPTER 10

TURNING FOES INTO FRIENDS

Despite the fact that poisonous plants can cause injury or death, humans have discovered through trial and error over millennia, and sometimes through serendipity or persistent scientific work, that many can be utilized for medicinal purposes even though they contain toxic substances. Other poisonous plants have found use as insecticides in the control of disease vectors and agricultural pests. In this final chapter, some of the compounds and the plants from which they can be derived are exemplified.

POISON TO MEDICINE: DISCOVERY OF ACTION

Throughout history plants have played a major role in human society. Even today, they are still feeding and clothing us, as well as providing building material and fuel for fires. As we have seen in previous chapters, they have also provided poisons for hunting, religious ceremonies and sources of medicines. The Swiss alchemist and physician Philippus Aureolus Theophrastus Bombastus von Hohenheim (1493–1541), better known as Paracelsus and often considered the father of toxicology, is known for his aphorism 'All things are poison and nothing is without poison; only the dose makes a thing not a poison.'

From a pharmacological point of view, herbal medicines are problematic. They always contain a mixture of compounds, some of which may be innocuous while others can increase the risk of poisoning or completely cancel the effect. The levels of active substances they contain varies depending on growing conditions, making it very difficult to dispense effective doses repeatedly from different batches of herbal material. As some of the substances used in medicine have a narrow therapeutic window (in other words, there is a fine line between a safe and therapeutic dose versus a harmful and poisonous amount), this has led to the development of modern drugs, usually containing isolated substances that can be accurately measured and dosed. Some plant-derived substances still play a major role as modern drugs after isolation of their active principle, and this is particularly true for substances that are too difficult or expensive to produce in a laboratory. In other cases, final production of a drug by chemists after isolation of precursors from the plant is necessary to prevent extinction of a species.

HELPING OR HURTING?

Although cardiac glycosides (see Chapter 3) can cause severe toxic effects at concentrations only slightly above the therapeutic range, the digitalis glycosides (see pages 198–199) have found a place in the treatment of congestive heart failure. Other herbal drugs with small therapeutic windows include opium (see pages 200–201),

LEFT **Engraving of Paracelsus,** who recognized that the difference between medicine and poison is the dose.

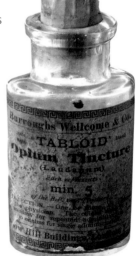

RIGHT **Laudanum, a dilute opium tincture,** was commonly taken as a painkiller in the eighteenth and nineteenth centuries, but many people became dependent on the medicine.

LEFT Calabar bean (*Physostigma venenosum*) seeds were used to make a drink called ésere that would be drunk during trial by ordeal (see page 207). The physostigmine they contain is used medicinally.

BELOW Himalayan mayapple (*Podophyllum hexandrum*) flowers are held above the young leaves. Podophyllotoxin, a compound with anticancer properties, is extracted from the rhizomes.

which contains effective painkillers but will cause dependence, and only a slight overdose of which may induce respiratory depression. For acetylsalicylic acid (see pages 202–203), the actual mechanism providing relief from fever and pain also adds to the risk of bleeding stomach ulcers. And the important antimalarial activity of *Cinchona* bark (see pages 204–205) is hampered by the fact that accumulation even after normal doses may lead to unwanted effects.

TRADITIONAL USE AND SERENDIPITY

Investigating the plants used in traditional medicinal systems has led to modern drugs that either reflect that traditional use or more surprisingly have quite different applications. The very harmful ordeal poisons of *Physostigma* and *Pilocarpus* species (see pages 206–207) have found use as antidotes and treatment for several diseases. The Madagascar periwinkle (*Catharanthus roseus*; see pages 208–209) was traditionally used to treat high blood sugar, and mayapples (*Podophyllum peltatum* and *P. hexandrum*; see pages 210–211) were laxatives and vermicides, but serendipitously they also turned out to be very good cancer drugs. Galantamine (see pages 212–213) was traditionally a paralysis treatment but has become a drug that slows down the advance of Alzheimer's disease. Other traditional drugs have retained more of their original use, such as the *Ephedra* alkaloids (see pages 214–215), used primarily to relieve respiratory cold symptoms, and several plants that are still used to control insect pests, including the neem tree (*Azadirachta indica*; see pages 216–217).

Heart of the matter – digitalis glycosides

Despite its potential lethality, purple foxglove (*Digitalis purpurea*) was used in European folk medicine to treat a number of ailments from at least medieval times. In Wales, ointments containing the plant were recommended for headaches and spasms by the famous physicians of Myddfai, and in England it was used against epilepsy, goitre and tuberculosis, and as an emetic. However, it was not until William Withering published a treaty on its use against dropsy in 1785 that purple foxglove became established as a drug with an acceptable risk profile.

An account of the foxglove

PLANT:
Digitalis purpurea L. and *D. lanata* Ehrh.
COMMON NAMES:
D. purpurea – purple (common) foxglove;
D. lanata – woolly foxglove
FAMILY:
plantain (Plantaginaceae)

TYPE OF TOXIN:
cardiac glycosides (cardenolides: digoxin, digitoxin)
SYMPTOMS OF POISONING IN HUMANS:
CIRCULATORY: arrhythmia, heart failure
NEUROLOGICAL: headache, weakness, confusion, coma
DIGESTIVE: nausea, vomiting, diarrhoea

BELOW **The English physician William Withering holding purple foxglove (*Digitalis purpurea*).**

Today, dropsy is recognized as the oedema resulting from congestive heart failure, where the heart muscle does not contract effectively, leading to leakage and accumulation of fluids between the body tissues. The therapeutic effects of some foxglove species observed as the alleviation of oedema are indirect through their action on the heart – by increasing the power of muscle contractions, less fluid will accumulate and thus hopefully enable the kidneys to remove the excess.

The English physician, botanist and chemist William Withering (1741–1799), considered by many to be the father of clinical pharmacology, published *An Account of the Foxglove and Some of its Medical Uses* in 1785. This scientific report examined 163 cases over ten years in which foxglove was administered. Through his careful study, Withering was able to establish that purple foxglove affected the contractions of the heart, and also deduced the most effective dose and dose intervals for treating oedema.

LEFT **Flowerhead of purple foxglove (*Digitalis purpurea*), also known as common foxglove, from western Europe and Morocco.**

BELOW **Digoxin has three units of the sugar digitoxose and is the most widely used cardiac glycoside in medicines.**

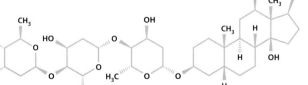

Digoxin

ABOVE **The eastern European perennial woolly foxglove (*Digitalis lanata*) has a higher concentration and greater diversity of digitalis glycosides than purple foxglove (*D. purpurea*) and is the main source of these for the pharmaceutical industry.**

DIGOXIN AND DIGITOXIN

As reported in Chapter 3, the active constituents of the foxglove genus (*Digitalis* spp.) are cardiac glycosides of the cardenolide type (see pages 54–59). Historically, crude foxglove plant extracts containing mixtures of the glycosides were used therapeutically. Of the many digitalis glycosides, the two most common in the treatment of congestive heart failure and atrial fibrillation (irregular, fast heartbeats) are digoxin and digitoxin. Two species of *Digitalis* have been important in the production of these drugs: purple foxglove, which is native to west, southwest and west-central Europe; and woolly foxglove (*D. lanata*), a native of eastern Europe. The purple foxglove is a biennial plant with a two-year lifespan, forming a rosette of leaves in the first year and a flowering stem in the second year. This means that it is a less suitable source for commercial production of cardioactive steroids than woolly foxglove, which is a perennial plant.

Digoxin and digitoxin are structurally similar compounds, differing only in a single hydroxyl group present in digoxin. This affects their elimination in the body: digitoxin is predominantly eliminated through metabolism in the liver and excreted very slowly, with an elimination half-life of five to nine days; while digoxin is mainly eliminated via the kidneys and has a half-life of 36–48 hours. The therapeutic dose range of digitalis glycosides is narrow and close to the toxic dose. Due to its long elimination time, digitoxin is used far less than digoxin, but the latter is contraindicated in patients with renal impairment, since this will rapidly lead to the accumulation of toxic concentrations of digoxin and consequently possible heart arrhythmias.

A more unusual historical use of purple foxglove in Ireland was in its alleged ability to identify a changeling – a fairy child left in place of a human child who had been stolen by fairies. If the child died after being given foxglove juice, then this was seen as evidence that it was indeed a fairy!

A dead gentleman of Verona

Cangrande della Scala (1291–1329) became ruler and lord of Verona in Italy in 1311, and is probably best known as a patron of the poet Dante Alighieri (c. 1265–1321). He was also a successful military campaigner and took control of several other Italian cities, but died at the age of 38 just four days after conquering Treviso. Rumours of poisoning started immediately. In 2004, della Scala's body was exhumed and toxic concentrations of digitalis glycosides were found in both his organs and in the content of his intestines, indicating possible foul play. We will never know if he was poisoned by his nephew and successor, or if his death was an inadvertent drug overdose.

In the arms of Morpheus – opium

Opium (from the Greek word *opos*, meaning 'juice') is the dried latex obtained from unripe capsules of the opium poppy (*Papaver somniferum*), in the poppy family (Papaveraceae; see pages 172–173). It is the greatest sedative-hypnotic drug in history. Along with its constituents and derivatives, opium has unparalleled efficacy in relieving pain and producing feelings of well-being and euphoria. However, according to the World Health Organization, 69,000 people die from opioid overdose each year.

The two faces of opium

PLANT:
Papaver somniferum L.
COMMON NAMES:
opium poppy
FAMILY:
poppy (Papaveraceae)

TYPE OF TOXIN:
benzylisoquinoline alkaloids of the morphinan type (morphine, codeine, thebaine)
SYMPTOMS OF POISONING IN HUMANS:
NEUROLOGICAL: constricted (pinpoint) pupils, drowsiness, coma, respiratory depression
DIGESTIVE: nausea, constipation

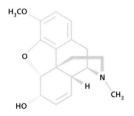

Codeine

ABOVE **Codeine**, or 3-*O*-methylmorphine, makes up about 2 per cent of the alkaloids in opium. In the human body it is metabolized by the liver into morphine, which is the active compound.

RIGHT **Opium poppy (*Papaver somniferum*)** is grown legally to supply the pharmaceutical industry for the manufacture of medical opioids, including codeine, morphine and derivatives.

The opium poppy has been cultivated and harnessed for its medicinal properties since antiquity. A number of artefacts from ancient Assyria depict clearly identifiable poppy capsules, and poppies feature in ancient Egyptian tomb paintings. The Ebers Papyrus, the earliest medical text of ancient Egypt (c. 1552–1534 BC), recommends an extract of poppy for children who cry incessantly. Its use continued into the nineteenth century, when laudanum, a tincture of opium, could be bought without a prescription. Although the opium poppy's addictive properties were well known, its analgesic effects have ensured repeated attempts to find safe ways to harness the positive side of this remarkable plant.

Morphine was the first alkaloid isolated from opium, in the early 1800s, and named after the god of dreams, Morpheus. Morphine could be produced in large quantities, and following the invention of the hypodermic syringe in the 1850s it was

RIGHT Opium is traditionally obtained by cutting the ripe capsule of the opium poppy (*Papaver somniferum*) and collecting the latex, an illegal practice in many countries.

used for managing post-operative and chronic pain, during minor surgical procedures and as an adjunct to general anaesthetics. In addition to its analgesic and narcotic properties, morphine is a powerful respiratory depressant and was previously used in cough elixirs. The side effect of constipation has also been utilised in anti-diarrhoeal medicines, as when morphine is mixed with kaolin.

However, like opium itself, morphine was found to be addictive, with a potential for abuse. In addition, its use is not very safe owing to these addictive effects, the euphoria it induces and its small therapeutic window. Much effort was consequently put into developing a non-addictive, safer and efficacious opiate, resulting in the semi-synthetic derivative diacetylmorphine (diamorphine), which although initially synthesized in 1874 did not become popular until it was resynthesized in 1898 and marketed under the name heroin. Unfortunately, this drug proved to be even more toxic and habit-forming than morphine, and is now widely used as an illegal recreational drug for its euphoric properties (see also pages 172–173).

Spongia somnifera

Arab recipes for early surgical anaesthetic preparations that date back to the ninth century list opium as a main ingredient, combined with henbane (*Hyoscyamus niger*) and mandrake (*Mandragora* spp.) in the potato family (Solanaceae; see pages 80–83 and 138–139). The preparation was administered using so-called somniferous or soporific sponges (*spongia somnifera* or *spongia soporifera*), made from fresh marine sponges that had been soaked in the liquid plant mixture and dried in the sun. When they were needed, the sponges were moistened with hot water before being applied to the patient's nostrils, the resulting effect being to render the patient unconscious. However, if too little was used the patient would not be sufficiently sedated, and if too much was used they would never wake up.

OTHER MEDICALLY IMPORTANT ALKALOIDS

Opium contains several isoquinoline alkaloids, including the addictive morphinans morphine and codeine, as well as the non-addictive thebaine. Opiates, such as morphine and codeine, affect the central nervous system (CNS) through stimulation of specific opioid receptors, which are widely distributed in the brain and also found in the spinal cord and digestive tract. These receptors respond to both endogenous transmitters (peptides produced by the body – so-called endorphins) and ingested plant alkaloids, which bind to them. Codeine is less potent than morphine, and is used as an analgesic for moderate to severe pain, such as migraine. Even so, problems with the chronic use of codeine are now becoming apparent, including increased tolerance and side effects such as abnormal pain sensitivity.

Thebaine, once considered an unwanted by-product of opium due to its lack of analgesic activity, has been used as the starting molecule for semi-synthesis of potentially useful new drugs that will be safer than morphine. In the 1960s, the chemist Kenneth Bentley attempted to achieve this by adding structures to thebaine. However, the so-called Bentley compounds were found to be highly addictive, in addition to having analgesic properties up to 12,000 times those of morphine, resulting in their alternative name of elephant morphines. In fact, one of them – etorphine (which is 5,000 to 10,000 times more potent than morphine) – is used in veterinary practice to sedate large animals such as elephants (*Loxodonta* spp. and *Elephas* spp.) and rhinoceros (*Rhinoceros* spp.).

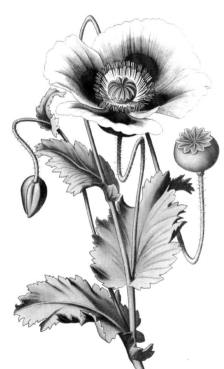

LEFT Opium poppy (*Papaver somniferum*) is an annual plant that can reach 1–1.5 m (3–5 ft) in height. In addition to being grown for the medicinal and illegal use of its alkaloids, it is also an attractive garden plant.

Plant panacea – salicylic acid

The Greek physician Hippocrates (c. 460–370 BC) was one of the first to record the use of salicylates for pain relief when he recommended chewing the leaves of willow in childbirth. Over the subsequent years, others described applications to reduce fever and inflammation, not only using willow but also other plants that contain similar compounds. In 1757, the Reverend Edward Stone started clinical trials on willow bark, leading to the isolation of the active compound and ultimately the development of the synthetic derivative aspirin (see box), which is easier on the stomach. Today, aspirin is one of the most widely used drugs in the world to relieve the symptoms of colds, flu, muscle pain, headaches, menstrual cramps and arthritis, and regular smaller doses are prescribed to reduce the risk of heart attacks and strokes.

INJECTING CHEMICAL PAIN

PLANT:
Salix species and *Filipendula ulmaria* (L.) Maxim. (syn. *Spiraea ulmaria* L.)

COMMON NAMES:
Salix spp. – willows;
F. ulmaria – meadowsweet

FAMILY:
Salix spp. – willow (Salicaceae); *F. ulmaria* – rose (Rosaceae)

TYPE OF TOXIN:
salicylates (*Salix* spp. – salicin; *F. ulmaria* – salicyladehyde)

SYMPTOMS OF POISONING IN HUMANS:
NEUROLOGICAL: dizziness, ringing in the ears, tremors, convulsions, coma
DIGESTIVE: abdominal pain, nausea, vomiting
OTHER: hyperpnea (increased rate and depth of breathing), liver damage with long-term use

LEFT **The glycoside salicin is a form of the active compound salicylic acid, which is itself present as an important hormone in all plants.**

RIGHT **The dried bark of species of willow, including white willow (*Salix alba*), is used medicinally to treat the fever and pain associated with a number of conditions and diseases.**

The Reverend Edward Stone (1702–1768) of Chipping Norton in England, aware of the folk use of the bark of white willow (*Salix alba*), tasted it and, finding it bitter, suspected that it might have the same properties as so-called Peruvian bark. This treatment for fever, also called quinine (*Cinchona* spp.; see pages 204–205), was widely used but expensive, so there was great interest in finding a cheap, local alternative. Stone spent the next six years undertaking clinical trials on himself and around 50 other people suffering from inflammatory disorders and agues (fevers). In 1763, he presented his promising results in the form of letter to the Earl of Macclesfield, president of the Royal Society. Although the general population was aware of the usefulness of willow, Stone's letter brought it to the attention of the medical community, giving rise to the modern history of salicylates.

Others confirmed Stone's findings, and salicin was finally isolated from willow bark in the late 1820s. A few years later, salicylaldehyde was distilled from meadowsweet (*Filipendula ulmaria*, syn. *Spiraea ulmaria*), and the active compound

ABOVE **White willow (*Salix alba*) is an upright tree from Europe, north Africa and temperate Asia whose male and female flowers are borne in separate heads known as catkins.**

salicylic acid was prepared from it. Salicylic acid was later isolated directly from salicin obtained from willow bark, and a third source, methylsalicylate from oil of wintergreen (*Gaultheria procumbens*) in the heather family (Ericaceae), followed. Salicylic acid is a plant hormone with roles in defence against pests and a variety of physiological processes, and is therefore widespread in many plant species. However, some plants such as willow accumulate high levels in the form of compounds called salicylates, such as salicin and methylsalicylate, which are converted to salicylic acid either when required by the plant or in the gastric tract of an animal that eats it.

A STRENGTH AND A WEAKNESS

Although salicylates are an effective treatment for pain, fever and inflammation, they also have a big downside in the form of severe gastric toxicity. This was explained in 1971 as the mechanism of aspirin's biological effects was discovered. By inhibiting enzymes called cyclooxygenases, salicylates halt the production of prostaglandins. These so-called local hormones regulate a number of physiological processes, such as pain perception, body temperature, the amount of protective mucus in the stomach lining and the platelets' ability to clump together and clot. Although aspirin taken in low doses prevents strokes in certain patients through its effects on platelets, higher doses diminish the mucus protection in the stomach and have been responsible for many deaths due to bleeding gastric ulcers.

The aspirin story

In 1893, Felix Hoffmann, a chemist working for Bayer in Elberfeld, Germany, was given the job of developing a salicylate that was gentler on the stomach (incidentally, he later went on to resynthesize diamorphine, which was then marketed as heroin; see pages 200–201). His search through the published literature revealed a salicylic acid derivative, acetylsalicylic acid, which had been isolated by the French chemist Charles Gerhardt in 1853. Hoffmann worked on an improved method of synthesizing acetylsalicylic acid, and successfully tested the compound on himself and his arthritic father, who had been intolerant of earlier salicylates. The new drug was given the trade name aspirin – 'a' for acetyl, 'spir' for *Spiraea*, and 'in', a popular drug name suffix – and announced in a paper in 1899, with no credit given to any of the people responsible for its discovery.

BELOW **Meadowsweet (*Filipendula ulmaria*) from Europe and temperate Asia is a perennial herbaceous plant that reaches 60–120 cm (2–4 ft) in height and bears irregular heads of flowers.**

Fever reliever – quinine

Cinchona trees (*Cinchona* spp.) are native to tropical mountainous regions of Central and South America, where a decoction of their bark is traditionally used to treat symptoms of recurring fever. This use was probably introduced to Europe by Jesuit priests (see box), but it soon became apparent that the bark was effective at treating only some forms of fever. In 1820, the active compound quinine was isolated, and some 60 years later the malaria parasite was first observed under a microscope. When the Second World War disrupted the global supply of cinchona bark (which by then was almost entirely being produced in Java), the development of synthetic alternatives effectively ended its pre-eminence in the treatment of malaria. The majority of the world's cinchona supply today is used by the soft drinks industry, although it still has a small but important medicinal role.

Quinine

LEFT **Quinine, one of the main quinoline alkaloids from cinchona bark, is often used as a bitter in tonic water.**

Cinchona and its alkaloids

PLANT: *Cinchona pubescens* Vahl and *C. calisaya* Wedd. (syn. *C. ledgeriana* (Howard) Bern. Moens ex Trimen)	**TYPE OF TOXIN:** quinoline alkaloids (quinine, quinidine, cinchonidine, cinchonine)
COMMON NAMES: cinchona, fever tree, Jesuit's bark, Peruvian bark, calisaya bark	**SYMPTOMS OF POISONING IN HUMANS:** **CIRCULATORY:** arrhythmia, hypotension, circulatory collapse, cardiac arrest
FAMILY: coffee (Rubiaceae)	**NEUROLOGICAL:** blurred vision, blindness, tinnitus, deafness
	DIGESTIVE: nausea, vomiting, abdominal pain, diarrhoea
	OTHER: headache, sweating

The history of the cinchona tree is reflected in its common names – fever tree, Jesuit's bark, Peruvian bark. However, its genus name, *Cinchona*, is an error – it was named after the Countess of Chinchón, who was treated for fever with the bark, so should have been spelt '*Chinchona*'. Only two species (*C. calisaya* and *C. pubescens*) among the 20 or so in the genus are main commercial sources of the active compounds, though many are used traditionally.

LEFT **The bark of *Cinchona calisaya*, an evergreen shrub or small tree native to tropical forests of Bolivia and Peru, is one of the main commercial sources of the antimalarial drug quinine.**

Quinine is the major alkaloid in *Cinchona* species and the best known, but other quinoline-containing alkaloids – such as quinidine, cinchonidine and cinchonine – are present in variable proportions. Prepared sulfates of quinine and the other three alkaloids were shown to be effective against malaria in one of the earliest large-scale clinical trials, conducted from 1866 to 1868 in 3,600 patients. All four alkaloids were found to result in greater than 98 per cent 'cessation of febrile paroxysms'.

The use of quinine has largely been superseded by other less toxic antimalarial medicines, such as the synthetic chloroquine, and the Nobel Prize-awarded natural product artemisinin from sweet wormwood (*Artemisia annua*) in the daisy family (Asteraceae). However, due to development of drug resistance by the malaria parasite, quinine is seeing a small resurgence in the treatment of severe cases of the disease.

Quinidine, a stereoisomer of quinine, is today used for cardiac complications, including arrhythmias, atrial flutter and fibrillation. A few countries also permit quinine or quinidine to be prescribed as a skeletal muscle relaxant for the treatment of 'night cramps', but there are concerns over the long-term use of these compounds.

Cinchonism and the bitter tonic

Gin and tonic was the drink of choice for officials stationed in the colonial outposts of Asia and Africa during the British Empire. Quinine was the bitter flavouring agent in tonic water, which was drunk as a prophylactic against malaria. Tonic waters around the world still contain quinine today, but the concentrations are too low to provide any antimalarial benefit or to give any concerns over their safety.

Chronic use or overdose of quinine, quinidine or cinchona bark can lead to a condition known as cinchonism, which is characterized by headache, tinnitus, abdominal pain, rashes, unusual bleeding/bruising and visual disturbances. Large doses of quinine can result in more serious symptoms of cinchonism, which include nerve damage leading to deafness and blindness, heart arrhythmia and death from cardiotoxicity. Except in severe cases, most symptoms of cinchonism are reversible if use of the drug involved ceases.

ABOVE **Cinchona (*Cinchona* sp.) bark drying in front of a longhouse in Sarawak, Southeast Asia. Cinchona trees have been cultivated outside their native Americas since the late nineteenth century.**

LEFT **Glass bottle that contained quinine sulfate, manufactured in the United Kingdom by Burroughs Wellcome and Co. in around 1860–1910, when the drug was an important treatment for malaria.**

Jesuit bark

Jesuit missionaries to the New World in the sixteenth century began to establish pharmacies – similar to European apothecaries – in areas they had come to evangelize. At first, they imported remedies from Europe, but they also began to research and obtain supplies of local medicines. The priests observed the traditional use of cinchona bark by the indigenous Quechua people of Peru, noting that they used decoctions of the bark to suppress shivering, one of the symptoms of fevers and malaria (which was likely introduced to South America via African slaves taken to the continent). The Jesuits purportedly introduced the dried bark powder to Europe by the 1630s. In 1677, cinchona bark first appeared in the London Pharmacopoeia as a treatment for fever or ague (malaria).

Anticholinergic antidote – physostigmine

The Calabar bean (*Physostigma venenosum*) is a climbing perennial plant in the legume family (Fabaceae) that grows along the banks of rivers in tropical Africa. The large, elongated fruit it produces contains two or three dark brown seeds (beans) that float, allowing effective dispersal by water. Several potent toxins are found within the bean, deterring any fish from eating them as they float downriver. Humans found a use for them as an ordeal poison, and more recently as an important antidote for tropane alkaloid poisoning and a treatment for myasthenia gravis.

One man's antidote is another man's poison

PLANT:
Physostigma venenosum Balf.
COMMON NAMES:
Calabar bean, ordeal bean
FAMILY:
legume (Fabaceae)
TYPE OF TOXIN:
physostigmine

SYMPTOMS OF POISONING IN HUMANS:
CIRCULATORY: faint, slow pulse
NEUROLOGICAL: dizziness, faintness, pinpoint pupils, convulsions
DIGESTIVE: increased salivation, involuntary urination and defecation
SKIN: increased sweating
OTHER: difficulty breathing

Physostigmine

LEFT **Physostigmine is an indole alkaloid also known as eserine. It inhibits the enzyme acetylcholinesterase.**

Toxicologists in the nineteenth century were intrigued by Calabar beans they had been sent from Nigeria. The effects on the body of the active compound physostigmine (eserine), isolated from the beans in the 1860s, seemed to be the exact opposite of those produced by atropine and similar tropane alkaloids (see pages 80–83), and it was therefore suggested that they might be an effective antidote to the other's poison. In 1871, 45 children and one adult became ill after eating beans that had been dropped during the unloading of cargo at Liverpool docks. The beans were Calabar beans and the prompt use of atropine as an antidote saved all but one of them. It is now known that physostigmine is not a true antagonist for atropine, but may still be a good antidote because it passes the blood–brain barrier rapidly and also because the two chemicals exert their opposite effects by interacting at different points of nerve signal transmission.

The Killer Bean of Old Calabar

In the 1840s, Scottish missionaries in the region of Nigeria known as Calabar discovered a practice of trial by ordeal. Calabar beans were given to those accused of serious crimes, such as murder, rape and witchcraft. The guilty would be poisoned and die, the innocent would suffer no ill effects and walk free. Locals trusted the method implicitly and hundreds would willingly swallow the beans to prove their innocence. The beans were even used as weapons for duels. The challenger would bite off half the bean and the opponent would eat the rest. The pair would continue in this way until one, or probably both, died.

Missionaries curious about the extraordinary properties of the beans sent samples back to Scotland where toxicologists discovered its medical potential. But how could the beans distinguish the guilty from the innocent? One theory has it that the innocent would confidently swallow the beans as quickly as possible to prove themselves. The guilty would hesitate and chew the beans before swallowing, trying to delay the inevitable result, but in fact giving themselves a much greater exposure to the toxins (see photograph of Calabar beans on page 197).

MEDICAL BENEFITS

Physostigmine blocks acetylcholinesterase, the enzyme responsible for breaking down the nerve-activating transmitter substance acetylcholine. A higher concentration of acetylcholine leads to nerve overstimulation, which can be disastrous and is the reason for poisoning by physostigmine. However, in the case of myasthenia gravis, an autoimmune disease causing muscle weakness, the acetylcholine receptors do not respond efficiently. Here, physostigmine, and especially its derivative neostigmine, can be used to prevent the breakdown of acetylcholine and thereby increase the amount available to stimulate the nerves. Another derivative of physostigmine, rivastigmine, is used for the symptomatic treatment of Alzheimer's disease.

Pilocarpine, isolated from the leaves of jaborandi (*Pilocarpus* spp.) in the citrus family (Rutaceae), is a direct antagonist for the effects of atropine – it interacts with the same acetylcholine receptors as atropine, but stimulates them rather than switching them off. The resulting effects of pilocarpine on the body are much the same as with physostigmine, as they both predominantly affect the nerves that control the body's 'rest and digest' response, and they therefore have some similar medical applications.

Pilocarpine and physostigmine are used to treat the symptoms of glaucoma (specifically the build-up of fluid in the eye that exerts pressure on the optic nerve) by causing the pupils to contract, thereby allowing fluid to drain more effectively from the interior of the eye. Other applications include treating a dry mouth, common after radiation therapy for head and neck cancers, although adverse effects from pilocarpine can include confusion, hallucinosis and agitation, even at normal doses.

LEFT The Calabar bean (*Physostigma venenosum*) from western central Africa is a woody climber that can reach 15 m (50 ft) in height and bears long pods containing two or three large seeds.

RIGHT Paraguay jaborandi (*Pilocarpus pennatifolius*) is a shrub or small tree 3–5 m (10–16 ft) in height with aromatic leaves. It contains low levels of pilocarpine, although the main commercial source of the medicine is Maranham jaborandi (*P. microphyllus*) from Brazil.

Anticancer agents – vinca alkaloids

The Madagascar periwinkle is native to the rainforest of the south and southeastern part of the island after which it is named. Somehow, several centuries ago this little plant managed to spread from its island home to the rest of world, where it is now cultivated as a tender ornamental and has become naturalized throughout much of the tropics and subtropics. Besides being grown as a beautiful garden flower, the Madagascar periwinkle was incorporated into many of the traditional medicinal systems in countries that adopted the species. Then, in the 1950s, it changed the outcome for many children suffering from leukaemia.

Diabetic lead

PLANT:
Catharanthus roseus (L.) G. Don (syn. *Vinca rosea* L.)
COMMON NAMES:
Madagascar periwinkle, rosy periwinkle
FAMILY:
dogbane (Apocynaceae)

TYPE OF TOXIN:
vinca alkaloids (especially vincristine and vinblastine)
SYMPTOMS OF POISONING IN HUMANS:
DIGESTIVE: nausea, vomiting, constipation, ileus
OTHER: immunosuppression

Madagascar periwinkle is used to treat a number of ailments in different traditional medicinal systems. In the 1950s, its use in the treatment of type 2 diabetes in Jamaica attracted the attention of Robert Noble (1910–1990) and Charles Beer (1915–2010), scientists at the University of Western Ontario in Canada, who were investigating its potential hypoglycaemic activity. They found that the plant extracts did not have an effect on blood sugar levels, but instead the test animals succumbed to bacterial infection as a result of depleted white blood cells. This result was considered serendipitous as the selective action suggested an anticancer potential for the plant, and an exhaustive search for the active constituents was initiated. In a parallel effort, the pharmaceutical company Eli Lilly was investigating 1,500 plant extracts in collaboration with the newly established Developmental Therapeutics Program at the National Cancer Institute in the United States (see box). Among the plants was Madagascar periwinkle, selected because it was used against diabetes in the Philippines.

Improving leukaemia survival

More than 150 alkaloids have been characterized from the Madagascar periwinkle to date. They are terpenoid indole alkaloids and many are known in other plants in the dogbane family. Anti-tumour activity has been demonstrated for a number of dimeric indole (bisindole) alkaloids found in Madagascar periwinkle, the most important being vincristine and vinblastine. These vinca alkaloids are so called because

Vincristine

LEFT **Vincristine is a dimeric monoterpene indole alkaloid that has become important as a drug against cancer.**

LEFT **Madagascar periwinkle (*Catharanthus roseus*) has opposite pairs of leaves and star-like pink or white flowers.**

the Swedish botanist Carl Linnaeus placed Madagascar periwinkle in the genus *Vinca* when he described it in 1759, and the use of that name has persisted despite botanists transferring the species to a new genus.

Sharing a mechanism of action with other natural products – including paclitaxel (see page 46), colchicine (see pages 152–153) and podophyllotoxin (see pages 210–211) – vinca alkaloids effect anticancer activity through their inhibition of cell division. They bind to the protein tubulin, preventing formation of microtubules, without which the cancer cells are unable to divide. Only minute amounts of vinblastine and vincristine occur in the plant material – more than 500 kg (1,100 lb) is needed to produce a single gram (0.03 oz) of vincristine. Fortunately, it is possible to produce these compounds semi-synthetically from simpler alkaloids that are found in the plant in much greater quantities.

Leukaemia is the most common childhood cancer, accounting for nearly a third of all cases. Thirty years ago, only one in two children with the most prevalent form of the disease, acute lymphoblastic leukaemia (ALL), survived. Children diagnosed with leukaemia today are four times as likely to survive for at least ten years than those in the 1970s, in large part due to the use of vincristine. The overall five-year survival rate for childhood leukaemia is currently 88 per cent, and for ALL survival this figure has increased to 92 per cent.

ABOVE **The Madagascar periwinkle (*Catharanthus roseus*) is cultivated in many countries, including its native Madagascar. Its leaves and flowers are harvested for the extraction of vinca alkaloids.**

Needle in a haystack

The discovery of the anticancer vinca alkaloids and podophyllotoxin-type lignans from mayapple (see pages 210–211) drove the National Cancer Institute (NCI), a United States government agency, to partner with the United States Department of Agriculture to collect plant material from thousands of species in their search for new anticancer drugs. Prior to the 1960s, their screening programme had focused on natural products created by fermentation and microbes. Thousands of plant extracts have since been assessed against a number of tumour types using high throughput screening by the NCI, leading to the development of paclitaxel (originally from the Pacific yew, *Taxus brevifolia*; see page 46), used to treat breast and ovarian cancers.

Warts and all – podophyllotoxin

Mayapple is an herbaceous plant in the barberry family (Berberidaceae) that grows in the woodlands of central and eastern United States. Its closest relative is a species in eastern Asia, Himalayan mayapple, which is sometimes placed in its own genus, *Sinopodophyllum*. Meaning 'foot leaf', the podophyllums are perennials with creeping rhizomes from which long-stemmed, lobed leaves arise. A resin from the rhizomes has been used traditionally in both North American and Asian systems of medicine for a variety of uses, but it is the external application to treat warts that has made its way into western medicine. Extracts of the resin are also proving useful for the synthesis of new anticancer compounds.

OLD USES OF *PODOPHYLLUM*

PLANT:
Podophyllum peltatum L. and *P. hexandrum* Royle (syn. *P. emodi* Wall. ex Honigberger, *Sinopodophyllum hexandrum* (Royle) T.S.Ying)

COMMON NAMES:
P. peltatum – mayapple, American mandrake;
P. hexandrum – Himalayan mayapple, Indian mayapple

FAMILY:
barberry (Berberidaceae)

TYPE OF TOXIN:
podophyllotoxin

SYMPTOMS OF POISONING IN HUMANS:
CIRCULATORY: hypotension
NEUROLOGICAL: depression of the CNS, peripheral neuropathy
DIGESTIVE: enteritis, vomiting, diarrhoea
SKIN: irritation, chemical burn (from topical use)

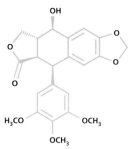

Podophyllotoxin

LEFT **The resin podophyllin contains the lignan compound podophyllotoxin, used in the treatment of genital warts.**

The fruit of the mayapple is edible, if insipid, but the rest of the plant is poisonous. However, despite its alternative common name of American mandrake, the species is not related to the true mandrake, which has a very different chemical profile (containing tropane alkaloids; see pages 80–83). Native Americans traditionally used a decoction of powdered mayapple rhizome and root, administered internally, as an anthelmintic (to remove parasitic worms), as an emetic (to induce vomiting) and as a laxative – a practice not without its

RIGHT **The mayapple (*Podophyllum peltatum*) spreads by rhizomes, from which stems arise at intervals. These bear a pair of leaves with a single flower in their axil.**

Sustainability of medicinal plants

The majority of medicinal plants traded globally today are harvested from the wild rather than cultivated. Habitat loss, combined with an increasing demand for herbal medicines, is raising concerns over the sustainability of this practice. As plants become rarer, their value as a commodity increases, causing additional pressure on wild populations such that their conservation becomes an issue.

The international community addresses this situation through the Convention on International Trade in Endangered Species of Wild Fauna and Flora (CITES). Along with large mammals such as the tiger (*Panthera tigris*), a number of medicinal plants are now listed under CITES, including Himalayan mayapple, which is recognized as one of the plant species currently threatened from overexploitation. CITES prohibits the international trade of all parts of the plant, excluding seeds and pollen, and finished products. Export from India of Himalayan mayapple and its products has been banned since 1984, but illegal trade is still believed to occur.

RIGHT The Himalayan mayapple (*Podophyllum hexandrum*) grows to 20–50 cm (8–20 in) in height, and has fleshy fruit that ripen to red and contain numerous seeds.

dangers. It was also applied externally as a poultice to treat warts and skin growths, as well as ulcers and sores.

The Himalayan or Indian mayapple, native to the Himalayas and countries bordering the mountain range, including Afghanistan and western China, has similar traditional medicinal uses to the North American species. Its rhizomes have been used as a purgative, laxative and anti-rheumatic.

Podophyllin resin and constituents

Extracts of dried rhizomes from both species of *Podophyllum* yield a resin known as podophyllin, in which the active compounds are lignans. The Himalayan mayapple rhizome produces more of the resin than the North American species (about 12 per cent resin, containing about 50–60 per cent lignans, compared with a yield of 2–8 per cent resin, containing 14–18 per cent lignans in mayapple). Podophyllin is included in some prescription-only medicines for the topical treatment of warts, such as genital warts caused by the human papilloma virus. The cytotoxic lignans – the most important of which is podophyllotoxin – are, however, too toxic to be used internally.

Anticancer activity

Podophyllotoxin was first isolated in 1880 and its chemical structure was elucidated in 1932. Although not unique to species of mayapple, the compound is present in other plants in only small amounts. In the 1940s, it was discovered that, like colchicine (see pages 152–153), podophyllotoxin binds to tubulin, preventing microtubule formation and thereby stopping cells from dividing. This mechanism of action is the basis of a number of drugs, including the vinca alkaloids (see pages 208–209) and paclitaxel (derived from the Pacific yew; see page 46), which are used to treat a range of cancers.

Although podophyllotoxin is active against tumours, it was found to be unsuitable for clinical use due to its unacceptable gastrointestinal toxicity. However, the compound is important in the manufacture of a number of semi-synthetic derivatives that have excellent anticancer activity and are far less toxic. These include etoposide, used against small cell lung cancer, and teniposide, used in the treatment of brain tumours.

Dementia drug – galantamine

With dementia, and particularly Alzheimer's disease, affecting 47.5 million people worldwide, new drugs to prevent or treat the symptoms are a priority for research. Two of the five drugs developed so far for the symptomatic treatment of Alzheimer's disease are derived from plants, one of which, galantamine, is found in members of the amaryllis family (Amaryllidaceae), including snowdrops (*Galanthus* spp.), daffodils (*Narcissus* spp.) and snowflakes (*Leucojum* spp.).

From polio to Alzheimer's

PLANT:
Narcissus pseudonarcissus L.
COMMON NAMES:
daffodil
FAMILY:
amaryllis (Amaryllidaceae)
TYPE OF TOXIN:
alkaloid (galantamine)

SYMPTOMS OF POISONING IN HUMANS:
CIRCULATORY: slow heart rate, arrhythmia
NEUROLOGICAL: dizziness, seizures (convulsions)
DIGESTIVE: nausea, vomiting, increased bowel movement
OTHER: sweating, increased urination, muscle weakness or spasms

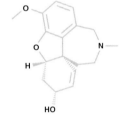

Galantamine

LEFT **Galantamine is a modified isoquinoline alkaloid present in many species of the amaryllis family, and is used to slow the progression of symptoms of Alzheimer's disease.**

The story of the development of galantamine (also known as galanthamine) as a licensed medicine began in the early 1950s, when a Russian pharmacologist was said to have noticed that people living at the foot of the Ural Mountains in the Caucasus were using local wild snowdrops, called Woronow's snowdrop (*Galanthus woronowii*), to treat poliomyelitis in children. Galantamine was isolated from *G. woronowii* in 1952, and it became a commercial product in Bulgaria a few years later for use in treating post-polio paralysis and neuromuscular pain.

The alkaloid galantamine occurs within the leaves and bulbs of various plants of the amaryllis family, and it is thought to protect them from herbivores and microbial infection. Pharmacological studies showed that it exhibits anticholinesterase activity, inhibiting the enzyme that degrades the neurotransmitter acetylcholine in the brain, thereby helping to maintain normal levels of acetylcholine. Since galantamine readily crosses the blood–brain barrier and also has beneficial effects by stimulating nicotinic receptors, it was selected as an ideal potential candidate in the treatment of Alzheimer's disease.

LEFT **Galantamine was first isolated from Woronow's snowdrop (*Galanthus woronowii*), found in the Caucasus and Turkey, but is now known to occur widely in the amaryllis family.**

LEFT The daffodil cultivar *Narcissus pseudonarcissus* 'Carlton' is an important source of galantamine. It is a widely grown garden plant that was bred in the United Kingdom by Percival D. Williams sometime before 1927.

Clinical trials on galantamine began in the 1990s and have consistently shown symptomatic benefit to cognition and clinical measures of Alzheimer's disease over several years of treatment, although the size of the effect is small. The alkaloid has been licensed as a medicine to treat Alzheimer's disease in the United Kingdom since the year 2000, and is also approved for use in the United States and many European and Asian countries. Another plant-derived drug to treat dementia symptoms is rivastigmine, which is chemically similar to physostigmine, an alkaloid found in the Calabar bean (see pages 206–207).

Drug sources

Although galantamine can be chemically synthesized, the daffodil (*Narcissus pseuodonarcissus*) and in particular its cultivar 'Carlton', is a major source of commercially produced quantities of the compound in western and central Europe. Research has shown that plants growing at higher altitudes produce far more galantamine than the same varieties grown at sea-level. The United Kingdom is the largest grower of daffodils in the world, with more than 4,000 ha (10,000 acres) put over to their commercial cultivation, much of it in the Black Mountains, near Brecon in Wales.

In parts of Europe, galantamine is extracted from wild populations of summer snowflake or Loddon lily (*Leucojum aestivum*). As 1 kg (2.2 lb) of bulbs is required to produce 1 g (0.03 oz) of galantamine, this is leading to a depletion of the natural populations. Due to the sustainability issue and increasing global demand, research in recent years has focused on developing new production methods involving cloned cell cultures, especially of the summer snowflake. In China, galantamine is predominantly obtained from cultivated red spider lilies (*Lycoris radiata*), while *Ungernia victoris* is the source in Uzbekistan and Kazakhstan.

Moly and Narcissus

The ancient Greek poet Homer was one of the first to write about the use of a plant as a protective agent. In his *Odyssey*, written in about the eighth century BC, Odysseus, following the advice of Hermes, uses a herb called moly as an antidote to Circe's magical potion, with which the enchantress had turned the hero's companions into pigs. The botanist Carl Linnaeus used the name for the species *Allium moly*, golden garlic, which has a yellow flower, but Homer's description that 'The root was black, while the flower was as white as milk' has led to more recent speculation that moly was a snowdrop (*Galanthus nivalis* or another species).

Another plant in the amaryllis family gets its scientific name from a character in Greek mythology. The young man Narcissus was known for his beauty but spurned all who approached. When walking beside water one day, he became entranced by his own reflection and, forgetting everything else, died of sorrow for the unattainable on the banks of the river or lake.

RIGHT The common snowdrop (*Galanthus nivalis*) from Continental Europe may have been the plant Homer called moly.

Stimulating interest – ephedra alkaloids

The drug *ma huang* has been one of the most important in the Chinese *materia medica* for millennia, having been first recorded in *Shen-nung Pen-ts'ao Ching* (*Divine Husbandman's Materia Medica*), the earliest extant Chinese pharmacopoeia (c. AD 25–200). Today, it is sourced from four species of ephedra (*Ephedra* spp.), including *E. sinica,* which is native to northeast China, Mongolia and parts of Russia. In fact, at least two-thirds of *Ephedra* species from across the globe are used medicinally, and are the source of useful compounds in the fight against cold and flu symptoms. Unfortunately, however, they are also subject to abuse.

An ancient drug

PLANT:
Ephedra sinica Stapf and other species
COMMON NAMES:
Chinese ephedra, desert tea
FAMILY:
ephedra (Ephedraceae)
TYPE OF TOXIN:
alkaloids (ephedrine, pseudoephedrine, norephedrine)

SYMPTOMS OF POISONING IN HUMANS:
CIRCULATORY: hypertension, arrhythmia, tachycardia
NEUROLOGICAL: dilated pupils, restlessness, irritability, insomnia, headache, seizures, stroke
DIGESTIVE: loss of appetite, nausea, vomiting
OTHER: urinary retention, difficulty breathing

BELOW **Chinese ephedra (*Ephedra sinica*) is a gymnosperm and close relative of conifers, including yews (*Taxus* spp.). The stems are photosynthetic and the fruit are fleshy cone bracts.**

The small genus of *Ephedra* contains around 54 species within its own family, Ephedraceae, distributed in northern temperate regions of the world as well as in western South America. There is evidence that ephedra may be one of the first plants that was used medicinally, as pollen of medicinal plants, including high-climbing jointfir (*E. altissima*), was found in the grave of a male Neanderthal buried in Shanidar Cave, Iraq, in around 60,000 BC. Some scientists dispute the interpretation that flowers had been placed deliberately in the grave, however, as the pollen may have been introduced by burrowing rodents.

Ephedrine

LEFT **Ephedrine is an alkaloid that resembles adrenaline both in structure and its stimulating activity in the human body.**

Indisputable recorded uses of ephedra in traditional medicine systems include the treatment of asthma, hay fever and other allergies, as well as respiratory diseases such as bronchitis, emphysema, and colds and influenza. The effectiveness of ephedra in treating many of these conditions is not in doubt, but abuse of the active compounds has required their use today to be controlled.

Ephedrine to amphetamine

Ephedra species contain several alkaloids: ephedrine, pseudoephedrine, norephedrine, norpseudoephedrine (cathine), methylephedrine and methylpseudoephedrine. Levels vary tremendously between the species (the North American species Nevada ephedra (*E. nevadensis*) is apparently devoid of them) and also between plant parts, with the alkaloids concentrated in green stems and leaves, while fruits and roots have virtually none.

Ephedrine and related alkaloids stimulate the nervous system by mimicking the effects of compounds naturally produced by the body that bind to and activate receptors (endogenous agonists). They are potent stimulators of receptors that are targets for adrenaline (epinephrine) and noradrenaline (norepinephine), and responsible for the 'fight or flight' response. The effects of these drugs include constriction of blood vessels (vasoconstriction), raised blood pressure, increased heart rate, expansion of bronchial tubes (bronchodilation), which makes breathing easier, and increase in energy expenditure (thermogenesis).

The two major alkaloids found in *Ephedra* species, ephedrine and pseudoephedrine, have been used in decongestant medicines to treat coughs, colds and sinusitis. However, these alkaloids are structurally similar to synthetic amphetamines, with ephedrine differing from

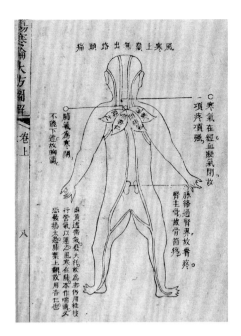

LEFT **The image and text of this woodcut illustration from 1833 give an account of the principles of diseases treated with a** *ma huang* **decoction, and the therapeutic effects of the remedy.**

methamphetamine only in a hydroxyl group, leading to the use of these medicines in the illicit manufacture of amphetamines. In response, restrictions are generally in place on the sale of products containing the alkaloids. In the United Kingdom, for example, they can legally be sold only at pharmacies, by or under the supervision of a pharmacist, with permitted levels of the alkaloids kept to a minimum if sold without a prescription.

Restrictions also apply to the sale of the herbal ephedra drug *ma huang* and others, which have been marketed as 'herbal ecstasy'. Despite these legal restrictions, the raw herb and products containing ephedra and its alkaloids are still openly sold over the Internet, posing a potential risk to consumers who are unaware of the extremely dangerous side effects.

BELOW **The Chinese herbal drug** *ma huang* **comprises the dried stems of any one of several species of** *Ephedra* **from China and India, although** *E. sinica* **is the most commonly used.**

Abusing the body beautiful

The stimulatory effects of ephedra and its alkaloids – including appetite suppression, weight loss and enhanced athletic performance – led to an increase in the popularity of supplements containing ephedra or ephedrine-type alkaloids (particularly in combination with other stimulants, including caffeine) among bodybuilders and those trying to lose weight rapidly. High doses of caffeine are thought to enhance the cardiovascular effect of ephedrine. However, the use of these diet and sports supplements has resulted in several fatalities due to heart attack and stroke, leading to their ban in the European Union and United States. The World Anti-Doping Agency has also prohibited the use of ephedrine and related alkaloids as stimulants. In addition to consumer abuse of these products, some manufacturers have been found to spike 'herbal' ephedra preparations with synthetic ephedrine alkaloids.

Nature's insecticide – azadirachtin

The neem tree is native to the Indian subcontinent and Southeast Asia. In Ayurvedic medicine, all parts of the tree are used to treat a variety of conditions, including skin diseases, malaria and intestinal worms. Traditionally, leaf and seed extracts are applied topically to treat infestations of lice, and dried neem leaves are used to protect clothes and food in storage from mould and insect pests. The observation that locusts leave neem trees untouched led to further study of the plant. Neem oil from the fruit and seeds, and the active compound azadirachtin, are now widely used as an ingredient in soaps and shampoos, and are one of the most important plant-based pesticides.

A NATURAL INSECTICIDE

PLANT:
Azadirachta indica A.Juss. (syn. *Melia azadirachta* L.) and *A. excelsa* (Jack) Jacobs

COMMON NAMES:
A. indica – neem, Indian neem; *A. excelsa* – marrango, Philippine neem

FAMILY:
mahogany (Meliaceae)

TYPE OF TOXIN:
triterpenoid limonoid (azadirachtin)

SYMPTOMS OF POISONING IN HUMANS:
CIRCULATORY: arrhythmias
NEUROLOGICAL: drowsiness, seizures, coma
DIGESTIVE: nausea, vomiting, abdominal pain
OTHER: liver failure

In 1952, the German entomologist Heinrich Schmutterer first reported the antifeedant effects of neem to the Western scientific community, following his observation that a swarm of desert locusts (*Schistocerca gregaria*) spared the neem trees yet devoured all other foliage in their path. Not only did this start an investigation of the activity of neem, but it also stimulated a wider interest in botanical insecticides as an alternative to synthetic compounds. The isolation in 1968 of the active compound, azadirachtin, from the seed kernels came at an opportune time – *Silent Spring*, written by Rachel Carson in 1962, had raised awareness of problems with persistent pesticides, which led to the ban of DDT in the United States in 1972.

LEFT Some birds eat the fruit of the neem tree (*Azadirachta indica*), including the Asian koel (*Eudynamys scolopaceus*) from woodland in tropical southern Asia – shown here is a female.

Azadirachtin

ABOVE Azadirachtin is a modified triterpene compound of the limonoid group, found in neem oil, and has become important as a natural antifeedant insecticide.

Neem, also known as Indian neem, is grown across the tropics and subtropics as a shade tree, for reforestation programmes and in plantations for production of azadirachtin, but is considered invasive in parts of Africa, the Middle East and Australia, where it has become naturalized. The seeds of Philippine neem (*Azadirachta excelsa*), which is native to Indonesia, Malaysia, Philippines, Papua New Guinea and Vietnam, and has naturalized in Singapore and Thailand, are also a source of neem insecticides. However, even though neem-based pesticides are a good biological alternative to synthetic compounds, accidental ingestion of neem products or seeds has resulted in a number of deaths, especially in children.

Effects on insects

While the effectiveness of neem insecticides is directly associated with azadirachtin content, the biological activity of many of the other compounds present in the neem tree (most of which are also triterpenoids of the limonoid group) add to its effect. Used in their natural combination, they may be helpful in mitigating the development of pesticide resistance.

The antifeedant activity of azadirachtin and some of the other neem compounds is through their stimulation of specific 'deterrent' cells on the insect mouthparts, while blocking other receptor cells that normally stimulate feeding, resulting in starvation and death of the insect. Insects vary considerably in their behavioural responses to azadirachtin. Studies on the desert locust have shown that it has a particularly high sensitivity to azadirachtin as an antifeedant, being deterred from feeding at concentrations of 0.04 parts per million. Interestingly, North American grasshoppers, including the

ABOVE **The neem tree (*Azadirachta indica*) can reach 10–20 m (30–65 ft) in height. It has compound leaves with several pairs of leaflets and heads of 150–250 small white flowers.**

American grasshopper (*Schistocerca americana*), which is in the same genus as the desert locust, are insensitive to azadirachtin at such low concentrations.

Insects that are not deterred from feeding on azadirachtin do not die immediately, but soon stop eating due to the action of the compound on a number of physiological pathways. It interferes with moulting and growth, for example, by blocking production and release of moulting hormones, causing moulting defects, and it disrupts reproduction by reducing the number of viable eggs and live progeny.

Other botanical insecticides

The Lithica poem dating from about 400 BC includes the phrase 'All the pests that out of earth arise, the earth itself the antidote supplies.' Until quite recently, humans have had to find natural ways of combatting disease and maximizing crop production. Botanical insecticides have played a part in this, and with the current move away from reliance on chemical insecticides their use is being re-examined. In addition to neem, nicotine from tobacco plants (*Nicotiana* spp.; see pages 98–99) in the potato family, rotenone from derris roots (*Derris* spp.) and other plants in the legume family, and pyrethrins from flowerheads of Dalmatian pyrethrum (*Tanacetum cinerariifolium*, syn. *Chrysanthemum cinerariifolium*) and Persian chrysanthemum (*T. coccineum*, syn. *C. coccineum*) in the daisy family have all played a major role globally.

LEFT **The flowers of the Persian chrysanthemum (*Tanacetum coccineum*), a perennial herbaceous plant native to the Caucasus, contain pyrethrins that are used as an insecticide.**

Glossary

acetylcholine nerotransmitter that binds to muscarinic and nicotinic receptors in the nervous system.

adrenaline neurotransmitter used in the nervous system; also known as epinephrine.

alkaloid nitrogen-containing compounds that often have pronounced physiological effects in humans.

allergenic can trigger an allergic reaction.

analgesic reduces or relieves the sensation of pain.

antagonist has an opposite effect to, or inhibits the activity of, another substance.

anticholinergic compound that inhibits effects of acetylcholine on both or either muscarinic or nicotinic receptors in the nervous system (cf. antimuscarinic).

antidote substance that counteracts the effects of a poison.

antifeedant discourages or stops feeding, particularly of insects.

antimuscarinic compound that inhibits effects of acetylcholine on muscarinic receptors in the nervous system. The receptors are so named because they are activated by the compound muscarine (cf. anticholinergic).

aril outer seed covering, often fleshy.

arrhythmia abnormal heart rhythm.

Ayurvedic medicine traditional system of medicine that originated in India.

barbiturate type of sleeping drug sometimes used to treat seizures.

binomial name consisting of two parts, such as the scientific names of plant species.

blood–brain barrier physiological barrier between the blood and cells in the brain and spinal cord.

carcinogenic having the potential to cause cancer.

central nervous system the brain and spinal cord in higher animals.

cultivar plant variety usually produced by selective breeding in cultivation.

cytotoxic compound that can cause cell death.

detoxification removing a toxic substance from a system, or chemically changing a substance to make it less or non-toxic.

dopamine neurotransmitter used in the nervous system.

dropsy older term for oedema.

emetic compound causing vomiting.

enzyme protein that can perform chemical reactions.

epinephrine see adrenaline.

glycoside non-sugar compound with one or more sugar compounds attached.

haemorrhagic causing bleeding.

hallucinogenic causing an altered mental state with visual or auditory sensations.

hepatotoxic having a harmful effect on the liver.

histamine signalling compound most associated with causing inflammation, especially due to allergies.

hypertension high blood pressure.

hypoglycaemia low level of glucose in the blood.

hypotension low blood pressure.

hypotonia abnormally low muscle tone; low osmotic pressure in a tissue.

ileus painful obstruction of the intestine.

lectin protein that can selectively bind different sugars.

mitochondrion (pl. mitochondria) cellular organelle responsible for energy production.

muscarinic receptor type of receptor that transmits nerve signals after activation, normally by acetylcholine but also by muscarine and similar compounds (cf. nicotinic receptor).

narcotic drug inducing an impaired mental state, pain relief and often euphoria.

nephrotoxic having a harmful effect on the kidneys.

neuron nerve cell.

neurotoxin having a harmful effect on nerve cells.

neurotransmitter substance released by nerve cells to send signals to other cells.

nicotinic receptor type of receptor that transmits nerve signals after activation, normally by acetylcholine but also by nicotine and similar compounds (cf. muscarinic receptor).

noradrenaline neurotransmitter used in the nervous system that can be converted to adrenaline by the addition of a methyl (CH3) group; also known as norepinephrine.

oedema swelling caused by the collection of fluid in body tissues or cavities.

opiate narcotic substance that can be isolated from opium.

opioid compound with the same effects as morphine.

ordeal poison compound used by humans to test the guilt or innocence of someone accused of a crime.

organelle structure within a cell that performs specific functions, e.g. mitochondrion.

osmosis process by which molecules of a solvent, such as water, pass over a semipermeable membrane from a less concentrated solution to a more concentrated solution.

osmotic pressure measure of the ability for a solution to cause osmosis.

parasympathetic nervous system the part of the human nervous system responsible for unconscious actions relating to 'rest and digest' and 'breed and feed' functions.

peripheral nervous system the nervous system outside the brain and spinal cord in higher animals, including nerves responsible for the five senses, voluntary movement and sending other signals between the body and brain.

phototoxic harmful in combination with ultraviolet radiation.

purgative laxative, usually with pronounced activity.

raphide needle-shaped crystal.

saponin plant-derived compound that foams in water due to having a water-soluble part (of sugars) attached to a lipid-soluble core structure.

sequestration an insect's ability to use ingested plant compounds for self-defence without chemical modification.

serotonin neurotransmitter used in the nervous system.

sodium channel proteins in cell membranes that allow selective passage of sodium ions.

steroid chemical structure found in compounds with several physiological activities, e.g. cardiac glycosides and sex hormones.

tachycardia rapid heartbeat; in adults at rest, more than 100 beats per minute.

taxonomy the science of naming, classifying and organizing organisms.

teratogenic able to damage the development of an embryo.

terpene compound derived from two or more isoprene units (for structure, see page 27).

toxin poisonous substance produced by a biological organism, e.g. a microbe, animal or plant.

trichome outgrowth from the surface of a leaf, stem, etc., such as a hair or bristle.

vacuole cellular organelle containing water solutions of inorganic and organic compounds, mainly for storage.

vasoconstriction the process of restricting the diameter of blood vessels.

Further reading

Elizabeth A. Dauncey, with toxicity by Leonard Hawkins and Katherine Kennedy, *Poisonous Plants: a guide for parents and childcare providers*. Royal Botanic Gardens, Kew, 2010.

Paul M. Dewick, *Medicinal Natural Products: a biosynthetic approach*. 3rd edition, Wiley, Chichester, 2009.

John Emsley, *Molecules of Murder: criminal molecules and classic cases*. RSC Publishing, Cambridge, 2008.

Dietrich Frohne and Hans Jürgen Pfänder, *Poisonous Plants: a handbook for pharmacists, doctors, toxicologists, biologists and veterinarians*. Manson Publishing, London, 2005.

James R. Hanson, *Chemistry in the Garden*. Royal Society of Chemistry, Cambridge, 2009.

Kathryn Harkup, *A is for Arsenic: the poisons of Agatha Christie*. Bloomsbury Publishing, London and New York, 2015.

Michael Radcliffe Lee, *Plants: healers & killers*. Royal Botanic Garden, Edinburgh, 2015.

John Robertson, *Is That Cat Dead? And other questions about poison plants*. Book Guild Publishing, Brighton, 2010.

John Harris Trestail III, *Criminal Poisoning: investigational guide for law enforcement, toxicologists, forensic scientists, and attorneys*. 2nd edition, Humana Press, New York, 2007.

Nancy J. Turner and Patrick von Aderkas, *The North American Guide to Common Poisonous Plants and Mushrooms: how to identify more than 300 toxic plants and mushrooms found in homes, gardens, and open spaces*. Timber Press, Portland and London, 2009.

Michael Wink and Ben-Erik van Wyk, *Mind-altering and Poisonous Plants of the World: a scientifically accurate guide to 1200 toxic and intoxicating plants*. Timber Press, Portland and London, 2008.

ONLINE RESOURCES

APGIV, Angiosperm Phylogeny Group version IV, a classification of flowering plants. Accessed via the Angiosperm Phylogeny website, www.mobot.org/MOBOT/research/APweb

Kew's Medicinal Plant Names Services portal, accessed via www.kew.org/mpns
Note that the scientific names of plants can change over time to reflect advances in understanding.

Other useful plant name resources can be found at: www.kew.org/science/data-and-resources/names

Index

abrin 146
Abrus precatorius 146
acerola family *see*
  Barbados cherry family
acetylcholine 38, 39,
  65, 92, 93, 99, 110,
  207, 212
acetylsalicylic acid
  197, 203
ackee 157, 158–9
Acokanthera schimperi 56
aconites 21, 46, 47,
  48–9, 124
aconitine 48, 49, 124
Aconitum spp. 21, 47,
  48–9, 124; *A. ferox* 15,
  48, 49; *A. lycoctonum*
  ssp. *vulparia* 48;
  A. napellus 21, 48;
  A. vulparia 48
Actaea racemosa 125
Adonis spp. 58, 124
adrenaline 35, 71, 215
Aethusa cynapium 101
Ageratina altissima
  161, 191
Aleurites fordii 115
alkaloids 27, 43, 64–5,
  75, 103, 124, 125, 135,
  152–3, 157, 172, 179,
  193, 213; aconite 46, 47,
  48–9, 124, 162;
  Amaryllidaceae 84–5,
  135, 142–3; benzo-
  phenanthridine 174–5;
  benzylisoquinoline 97,
  173, 200-1; bisbenzyl-
  isoquinoline 95; bisindole
  208; capsaicinoid 122–3,
  139; *beta*-carboline 86–
  7; diterpene 46–51, 124;
  Ephedra 197, 214–15;
  ergot 88–9, 191; erythrinan 97; glyco- 135, 138,
  139, 140–41; indole 52,
  65, 66–9, 86–7, 88, 96,
  206, 208–9; indolizidine
  190–91; isoquinoline 84,
  97, 136–7, 152–3, 173,
  201, 212; monoterpene
  indole 52, 65, 66–9, 96,
  136; phenethylisoquino-
  line 152–3; piperidine 92,
  98–101, 139, 152, 193;
  pyrrolizidine 34, 43, 157,
  161, 164–7, 190; quino-
  line 204–5; quinolizidine
  93, 104–5; *Taxus* 46–7,
  50–51; tetrahydroiso-
  quinoline monoterpene
  136–7; tropane 65,
  80–83, 93, 138, 206;
  vinca 208–9, 211
alkylcatechols/
  alkylresorcinols 130

Allium spp.: *A. cepa* 143,
  152; *A. moly* 213;
  A. sativum 143;
  A. schoenoprasum 143;
  A. ursinum 58, 152
almond 23, 183
aloe family
  (Asphodelaceae) 144
Aloe spp. 144;
  A. barbadensis 144;
  A. ferox 144; *A. vera* 144
Alzheimer's disease 197,
  207, 212–13
Amanita spp.: *A. phalloides*
  34, 161; *A. virosa* 34,
  35, 161
Amaryllis belladonna 142
amaryllis family (Amaryl-
  lidaceae) 84, 135, 142,
  143, 212, 213
amatoxins 34, 35
amino acids (non-protein)
  25, 26, 27, 102, 159,
  178, 186, 188
gamma-aminobutyric acid
  (GABA) 38, 39, 64, 70
Ammi majus 129
Amsinckia spp. 165
amygdalin 183
Anacardiaceae
  see cashew family
Anacardium occidentale
  130, 131
Anadenanthera spp.:
  A. colubrina 38;
  A. peregrina 65, 102
Anamirta cocculus 70, 71
Andromeda spp. 78
andromedotoxin 78
Anemone spp. 124, 126;
  A. hepatica 127
anemonin 126, 127
angel's trumpets 4, 21,
  41, 43, 65, 83, 138
anisatin 70
Anthoxanthum nitens 171
anthraquinones 135,
  144–5
Anthriscus sylvestris
  75, 129
Antiaris toxicaria 59
Apiaceae *see* carrot family
Apium graveolens 19, 129
Apocynaceae
  see dogbane family
Apocynum cannabinum 52
apples 183; bitter 151;
  devil's 82, 83; monkey 67
apricot 182, 183
Araceae *see* arum family
Argemone mexicana
  173, 174–5
Argyreia nervosa 89
Aristolochia spp. 35,
  157, 168, 169;
  A. clematitis 168–9;
  A. fangchi 168, 169;
  A. manshuriensis 169
Aristolochiaceae *see*
  birthwort family

aristolochic acids
  35, 157, 168–9
arrow poison 50, 52, 55,
  56–7, 59, 71, 84–5, 92,
  94–7, 114, 149
Artemisia spp.: *A. annua*
  205; *A. dracunculus* 160
artemisinin 205
arum family (Araceae)
  108, 112–13, 143
Asclepiadaceae 52
Asclepias spp.
  43, 52, 53
asparagus family (Aspar-
  agaceae) 54, 58, 60–61
asparagus pea 147
Asperula odorata 171
Asphodelaceae *see* aloe
  familyw
aspirin 202, 203
Asteraceae *see* daisy family
Astragalus spp. 190
atisine 49
atractylosides 156, 161,
  162–3
Atropa bella-donna
  15, 23, 80, 81, 138
atropine 23, 41, 65, 80,
  81, 82, 206, 207
autumn crocus family
  (Colchicaceae) 152–3
ayahuasca vine 65, 86–7
Azadirachta spp.:
  A. excelsa 217; *A. indica*
  197, 216, 217
azadirachtin 216–17
azalea 21

bakuchi 129
Balkan endemic
  nephropathy (BEN) 168
Banisteriopsis caapi 65,
  86–7
Barbados cherry family
  (Malpighiaceae) 86
barberry family
  (Berberidaceae) 125, 210
beakwort 127
bean family *see* legume
  family
bedstraw family *see*
  coffee family
belladonna 81
Bentley compounds 201
Berberidaceae *see*
  barberry family
berberine 173
bergamot orange 129
bindweed family *see*
  morning glory family
birthworts 157, 168–9
birthwort family
  (Aristolochiaceae) 168
bishop's weed 129
black cohosh 125
Blighia sapida 157, 158
blood lilies 142
bloodroot 172, 173
bog rosemary 78
boneset 165

Boophone spp 84, 85;
  B. disticha 65, 84–5
borage family
  (Boraginaceae)
  67, 164, 165, 166
bracken 179, 192–3
Brassica spp.: *B. juncea*
  121; *B. nigra* 121, 175;
  B. oleracea 120
Brassicaceae
  see cabbage family
Brazilian pepper tree 131
bread 89, 166, 169, 172
breadpalm 189
Brodie, Sir Benjamin 94
brooms 104; scotch 104
Brugmansia spp. 21, 41,
  43, 83, 138; *B. sanguinea*
  4, 83; *B. suaveolens* 65
buckthorn family
  (Rhamnaceae) 144, 187
buckthorns 144;
  alder 135
buckwheat 129
buckwheat family
  see dock family
bufadienolides 47, 54–5,
  58, 60–61, 124
buphanidrine 84–5
butter bean 147
butterburs 161
buttercups
  124, 126, 127; bur 126,
  127; creeping 124;
  curveseed 126
buttercup family
  (Ranunculaceae) 46, 47,
  48, 54, 58, 60, 109,
  124–5, 126–7

cabbage 120, 121
cabbage family
  (Brassicaceae) 109,
  120–21, 175
calabar bean 197, 206,
  207, 213
calcium oxalate crystals/
  raphides 18, 108,
  112–13, 143
Callilepis laureola 157, 162
Caltha spp. 126
camotillo 189
Camptotheca acuminata
  43
camptothecin 43
cannellini beans 147
caper family (Capparaceae)
  120
capers 120
Capparis spinosa 120
capsaicin 109, 116,
  121, 122–3
Capsicum spp. 122, 138,
  139; *C. annuum* 122,
  123; *C. chinense* 122;
  C. chinense 'Carolina
  Reaper' 123;
  C. frutescens 122
Carapichea ipecacuanha
  136–7

cardenolides 37, 43, 47,
  53, 54–5, 56–9, 60, 103,
  124, 198–9
cardiac/cardioactive glycosides 37, 46, 47, 54–61,
  124, 196, 199
Carica papaya 120
Carlina gummifera
  157, 161
carrot family (Apiaceae)
  74–5, 76, 100–01,
  109, 128–9
Cascabela thevetia 58
cashew 130, 131
cashew family
  (Anacardiaceae) 26, 130
cassava 115, 184–5, 187
Cassia spp. 102
castor oil plant
  114, 146, 148–9
Catharanthus roseus 43,
  52, 197, 208–9
cathine 215
celandine: lesser 125, 126,
  127; greater 173
celery 19, 76, 129
cephaeline 136–7
Cephaelis spp.:
  C. acuminata 136;
  C. ipecacuanha 136
Ceratocephala spp. 126,
  127; *C. falcata* 127;
  C. testiculata 126, 127
Cerbera spp. 53:
  C. manghas 53;
  C. odollam 37, 53
Cestrum spp. 139;
  C. laevigatum 163;
  C. parqui 163
chaconine 140
Chailletia toxicaria 180
chakruna/chacruna 87
*Chamaemelum
  nobile* 160
chamomile 160
charmac 166
Charybdis spp. 61;
  C. maritima 60
Chelidonium majus 173
chickling vetch 186
chilli peppers 109, 116,
  122–3, 138, 139
Chinese goldthread 125
*Chondrodendron tomento-
  sum* 92, 94, 95
Christmas rose 124, 126
Christmas vine 89
Chrysanthemum spp.:
  C. cinerariifolium 217;
  C. coccineum 217
Cicuta spp. 64, 76, 77;
  C. virosa 20, 74, 76, 77
cicutoxin 74, 76, 77
Cimicifuga racemosa 125
Cinchona spp. 197, 202,
  204–5; *C. calisaya* 204;
  C. ledgeriana 204;
  C. pubescens 204
cinchonine/cinchonidine
  204, 205

Cinnamomum spp.:
C. cassia 171;
C. verum 171
cinnamon 171
Citrullus spp.:
C. colocynthis 151;
C. lanatus 150
citrus family (Rutaceae) 109, 129, 207
Citrus spp. 129;
C. bergamia 129
Claviceps purpurea 89
Clavicipitaceae 88
Clematis spp. 21, 124, 126; *C. vitalba* 125
Clivia spp. 142
clover 103; sweet 102, 157, 170–71
Cnidoscolus spp. 111, 114;
C. angustidens 111;
C. megacanthus 111;
C. stimulosus 111;
C. texanus 111;
C. urens ssp. *stimulosus* 111
Cocculus indicus 71
cocklebur 139, 156, 162, 163
codeine 173, 200, 201
coffee family (Rubiaceae) 66, 69, 87, 137, 171, 181, 204
Colchicaceae *see* autumn crocus family
colchicine 135, 152–3, 209, 211
Colchicum autumnale 152–3
Colocasia spp.:
C. antiquorum 113;
C. esculenta 113
colocynth 151
colorin tree 97
colorines 97
coltsfoot 161
comfrey 157, 164, 165, 190
coniine 75, 100–01
Conium maculatum 75, 100
Convallaria spp. 58;
C. majalis 54
Convolvulaceae *see* morning glory family
coral tree 97, 102
coriamyrtin 72, 73
Coriaria spp. 64, 72;
C. arborea 72, 73;
C. japonica 72;
C. myrtifolia 72
Coriariaceae *see* tanner's bush family
Coronilla spp. 103
Corynanthe johimbe 69
coumarins 34, 157, 170–01
cow parsley 75, 129
cowbane 20, 74, 76–7
coyotillo 187
Crinum spp. 142

Crippen, Dr Hawley Harvey 138
crocus 152; autumn 152–3; saffron 152
Crocus spp. 152;
C. sativus 152
Crotalaria spp. 166
Croton spp. 114;
C. tiglium 118
cucumber family (Cucurbitaceae) 150–51
cucumbers 150; bitter 19, 150; squirting 150, 151; wild 150, 151
Cucumis spp.: *C. africanus* 150, 151; *C. sativus* 150
Cucurbitaceae *see* cucumber family
cucurbitacins 135, 150–51
Cullen corylifolium 129
curares 25, 67, 92, 94–7, 100, 102
curcin 114, 146
cyanide 35, 179, 180, 182–5
cyanobacteria 12, 188–9
cyanogenic glycosides 35, 43, 103, 115, 178, 179, 182–5, 188, 192
cycads 178–9, 188–9; sago 188–9
Cycadaceae 188
Cycas spp.: *C. media* 178, 179, 189; *C. micronesica* 189; *C. revoluta* 188, 189
cycasin 188
cyclopamine 193
Cynoglossum spp. 165
cytisine 102, 104, 105
Cytisus scoparius 104

daffodils 142–3, 212–13
da huang 144, 145
daisy family (Asteraceae) 125, 139, 156, 160–61, 162, 164, 166, 191, 205, 217
daoun setan 111
Daphne mezereum 119
daphnetoxin 119
Darling pea 190
Darwin, Charles 16
Datura spp. 83, 138;
D. stramonium 22, 83
Daucus carota 101
dead man's fingers 76
death cap 34, 161
Delphinium spp. 47, 48, 124, 190
Dendrocnide spp.:
D. gigas 111;
D. moroides 110, 111
Dermatophyllum spp. 104;
D. secundiflorum 104–5
Derris spp. 217
destroying angel 34, 35, 161
detoxification 34, 35, 49, 115, 120, 151, 157, 185, 187, 188

devil leaf 111
devildoer 96
Dichapetalaceae *see* gifblaar family
Dichapetalum spp.:
D. cymosum 180–81;
D. toxicarium 180
dicoumarol 157, 171
Dieffenbachia spp. 108, 112; *D. picta* 112;
D. seguine 108, 112
digitalis glycosides 196, 198–9
Digitalis spp. 55, 58, 199;
D. lanata 198, 199;
D. purpurea 46, 198, 199
digitoxin 198, 199
digoxin 55, 58, 198, 199
dihydrosanguinarine 174, 175
N,N-dimethyltryptamine (DMT) 65, 86, 87
Dipteryx odorata 170
dock family (Polygonaceae) 129, 144
doctrine of signatures 127, 168
dogbane family (Apocynaceae) 37, 52–3, 54, 56, 58, 66, 68, 69, 208
dopamine 38, 39, 65, 93
Drimia spp. 60, 61;
D. maritima 60
dumbcane 108, 112
Dutchman's pipe 168

eagle fern 192
Ecballium elaterium 150
Echium spp. 165;
E. vulgare 165
Elaeis guineensis 26
elephant vine 52
emetine 136–7
emodin 144, 145
Encephalartos longifolius 189
ephedra family (Ephedraceae) 214
Ephedra spp. 214–15;
E. altissima 214;
E. nevadensis 215;
E. sinica 214, 215
ephedrine 214–15
epidemic 119, 159, 187, 193; dropsy 175
epinephrine *see* adrenaline
ergot fungus 89
ergotamine 89
Ericaceae *see* heather family
erva-de-rato 181
Erythrina spp. 97, 102;
E. americana 97;
E. berteroana 97
etoposide 211
Eupatorium spp.:
E. cannabinum 161;
E. purpureum 161;
E. rugosum 191

Euphorbia spp. 18, 109, 114, 116, 117;
E. cotinifolia 111;
E. damarana 115;
E. pulcherrima 21, 117;
E. resinifera 116;
E. tirucalli 114, 117
Euphorbiaceae *see* spurge family
euphorbium 116

Fabaceae *see* legume family
fadang 189
fagopyrin 129
Fagopyrum esculentum 129
fever tree 204
Ficaria verna 125
fiddleneck 165
Filipendula ulmaria 202, 203
finger rot 111
fireball lily 85
fish berry 70, 71
fish poisons 27, 71, 114, 117, 119
flame lily 152, 153
fluoroacetic acid 178, 180–81
fluorocitric acid 180, 181
Foeniculum vulgare 101
fool's parsley 101
foxgloves 46, 55, 58, 198–9
Frangula spp. 144;
F. alnus 135
fumitory family (Fumariaceae) 172
Fumaria muralis 172
furanocoumarins 19, 75, 128–9

galantamine 197, 212–13
Galanthus spp. 142, 212;
G. nivalis 213;
G. woronowii 212
Galium odoratum 171
Gandhi, Mahatma 69
garden huckleberry 139
garlic 143; golden 213; wild 58, 152
Gastrolobium spp. 103, 181; *G. spinosum* 178
Gaulthecaceae 66, 67
gelsemine 67
Gelsemium spp. 67:
G. elegans 64, 67;
G. rankinii 67;
G. sempervirens 67
Genista spp. 104
Gerhardt, Charles 203
German primula 109
gifblaar 180, 181
gifblaar family (Dichapetalaceae) 180
Gloriosa superba 152, 153
glory lily 153

beta-glucosidase 126, 182
glucosinolates 120–21
Glycine max 42
golden rain/chain 93, 104, 105
goldenseal 125
grass family (Poaceae) 88, 171, 183
grass pea 102, 185, 186–7
grayanotoxins 65, 78–9
groundsel 161, 167
guan mu tong 168
guang fang ji 168, 169
gympie bush 110, 111

Haemanthus spp. 142
han fang ji 169
Haplopappus heterophyllus 191
happy tree 43
harmala 86
harmine 86, 87
hathisunda 165
Hawaiian baby woodrose 89
heartbreak grass 64, 67
heather family (Ericaceae) 65, 78, 203
heliotrope 165
Heliotropium spp.:
H. eichwaldii 165;
H. ellipticum 165;
H. popovii ssp. *gillianum* 166
hellebores 60, 124, 126; black 124; California false 193; white 43
Helleborus spp. 60, 124;
H. niger 124, 126
hemlock 75, 100–01
hemlock water dropwort 74, 76–7
hemp agrimony 161
henbane 81, 138, 201
Hepatica nobilis 127
Heracleum spp. 128;
H. mantegazzianum 75, 128–9;
H. sosnowskyi 128
Hierochloe odorata 171
Hippomane mancinella 114, 118
histamine 108, 110
Hoffmann, Felix 203
hogweeds 75, 128–9
holy grass 171
honey 73, 79, 164
hound's tongue 165
Hyaenanche globosa 71
Hydrastis canadensis 125
hyenanchin 71, 72, 73
hyoscine *see* scopoloamine
Hyoscyamus spp. 81, 138;
H. niger 81, 138, 201
hypericin 129
Hypericum spp. 129;
H. perforatum 129
hypoglycins 158–9

iboga 65, 68
ibogaine 68
Illicium spp. 70, 71:
  I. anisatum 64, 70;
  I. verum 70
impila 157, 162, 163
inkberry 163
insecticides 42, 93, 98, 99, 139, 161, 216–17
ipecac 136, 137
ipecacuanha 135, 136–7
Ipecacuanhae Radix 137
Ipomoea spp. 191;
  I. tricolor 88–9;
  I. violacea 89
Isocoma pluriflora 191
isoprene/isoprenoid 25, 27
isothiocyanates 109, 120–21; allyl 121

jaborandi 207
Jacobaea vulgaris 161, 166–7, 190
jacobine 166
Jamaica walnut 71
Jascalevich, Dr Mario 96
Jatropha spp. 114, 146:
  J. cathartica 114;
  J. curcas 114;
  J. gossypiifolia 111
Jersey lily 142
jessamines 67
Jesuit's bark 204, 205
jicamilla 114
jimsonweed 22, 83, 138
Joe Pye weed 161
Jussieu, Antoine Laurent de 14

Kalmia spp. 78;
  K. angustifolia 78;
  K. latifolia 78
Karwinskia humboldtiana 187
kidney beans 102, 146, 147, 186
Kirk, John 57
knitbone 165
kombe 55, 56–7
konzo 178, 182, 185, 187
kowhai 104

Laburnum spp.:
  L. alpinum 104;
  L. anagyroides 93, 104, 105
lactones: ring 55, 70; sesquiterpene 64, 70–73; steroidal 46; unsaturated 126–7
Landsteiner, Karl 147
Laportea moroides 110
larkspurs 47, 48, 124, 190
Lathyrus spp. 185;
  L. odoratus 187;
  L. sativus 102, 186–7
laudanum 196, 200
laurel: mountain 79; sheep 79; sheep kill 79; spurge 119; Texas mountain 104

laurel family (Lauraceae) 171
lectins 114, 135, 143, 146–9, 186
legume family (Fabaceae) 42, 86, 93, 97, 102–3, 104–5, 109, 144, 146–7, 164, 166, 170, 178, 181, 186, 190, 206, 217
Lent lily 142
leopard lily 112
Leucojum spp. 212;
  L. aestivum 213
leukaemia 119, 208–9
Levant berry 71
lignans 209, 210, 211
lily-of-the-valley 54, 58
lima bean 147
lime disease 129
linamarin 115, 184
Linaria vulgaris 16
Linnaea borealis 14
Linnaeus, Carl 14, 15, 16, 50, 61, 103, 166, 209, 213
Litchi chinensis 159
liverwort 127
locoweeds 179, 190–91
Loddon lily 213
Loganiaceae 66, 67, 94, 96
Lophophora williamsii 65, 105
lotaustralin 184
Lotus tetragonolobus 147
lucky beans 97
lupin 105
lupinine 105
Lupinus albus 105
lychee 159
Lycium chinense 138
lycorine 142–3
Lycoris spp. 142;
  L. radiata 142, 143, 213
lysergic acid diethylamide (LSD) 65, 88

ma huang 214, 215
ma sang 72
Madagascar periwinkle 43, 52, 197, 208–9
Maharbal 82
mahogany family (Meliaceae) 216–17
Maiwein 171
mala mujer 111
mallow family (Malvaceae) 191
Malpighiaceae *see* Barbados cherry family
Malus domestica 183
Malvaceae *see* mallow family
manchineel tree 114, 118
Mandragora spp. 81, 82, 138, 201; *M. autumnalis* 82; *M. officinarum* 82
mandrakes 81, 82, 138, 201, 210; American 210
Mangifera indica 130
mango 130

Manihot esculenta 115, 184–5
manioc 184
marsh marigold 126
mayapples 197, 209, 210–11
meadow saffron 152
meadowsweet 202, 203; Japanese 49
Melanthiaceae *see* trillium family
Melia azadirachta 216
Meliaceae *see* mahogany family
Melilotus spp. 102;
  M. albus 170;
  M. dentatus 171;
  M. officinalis 157, 170
Mendel, Gregor 17
Menispermaceae *see* moonseed family
Mentha spp. 122
menthol 122
Merian, Maria Sibylla 103
mescal beans 104, 105
mescal buttons 87
mescaline 65, 87, 105
beta-methylamino-L-alanine (BMAA) 178, 188, 189
mezereon 119
mezereum family (Thymelaeaceae) 116, 119
milk sickness 161, 179, 191
milkbushes: African 117; Damara 115
milkthistle 161
milkvetch 190
milkweeds 43, 52, 53
millet, great 183
mint 122
moita de calango 191
monkshood 21, 48
moonseed family (Menispermaceae) 71, 94, 95, 169
Moraceae *see* mulberry family
morning glory family (Convolvulaceae) 88, 89, 191
morphine 25, 27, 41, 173, 200–01
mulberry family (Moraceae) 59, 109
mustards 121, 175
mustard bomb 22, 120–21
mustard oil 109, 175
myrtle–leaved coriaria 72

naked lady 152
Narcissus spp. 142, 212;
  N. pseudonarcissus 142–3, 212–13;
  N. pseudonarcissus 'Carlton' 213
nasturtium 120
Natal lilies 142

nectar 21, 73, 79, 93, 157, 167
neem tree 197, 216–17
neonicotinoids 93
Nerium oleander 8, 12, 58
nettle family (Urticaceae) 108, 110–11
nettles 108, 110–11
Nicotiana spp. 98–9, 139, 217; *N. attenuata* 99; *N. glauca* 98; *N. rustica* 98; *N. tabacum* 92, 98
nicotine 12, 43, 92, 93, 98–9, 100, 104, 105, 217
nightshade family *see* potato family
nightshades 15, 138; black 138, 139; deadly 23, 81, 138; silverleaf 135; woody 139
Nitrariaceae 86
noradrenaline 38, 39, 65, 215
norephedrine 214–15
norepinephine *see* noradrenaline
noseburn, branched 111
Nostoc spp. 189
nux vomica 64, 66, 96

odollam tree 37, 53
oak, sessile 33
Oenanthe spp. 64, 76–7;
  O. crocata 74, 76;
  O. javanica 76
oenanthotoxin 74, 76–7
oleander 8, 12, 58; climbing 57; yellow 58
ololiuqui 89
onions 142, 143, 152; climbing 61; sea 60; spring 143
onion family 84
opiates 41, 68, 201
opium 173, 196, 200–01
ordeal bean 206; Madagascar 53
ouabain 56, 57, 59
oxeye daisy 162; creeping 163
oxytrope 190
Oxytropis spp. 190;
  O. lambertii 191;
  O. sericea 190

paclitaxel 43, 46, 50, 209, 211
Palicourea marcgravii 181
Papaver spp. 173;
  P. somniferum 18, 22, 27, 87, 172, 200–01
Papaveraceae *see* poppy family
papaya 120
Paracelsus 127, 196
pareira brava 92, 94, 95
parquin 163
parsnip 18, 74, 76, 101, 129

Pascalia glauca 163
pasqueflower 126
Pastinaca sativa 18, 101, 129
Pausinystalia johimbe 69
pea family *see* legume family
peach 23, 183
Peganum harmala 86
pencil tree/cactus 114, 117
Persian chrysanthemum 217
Peruvian bark 202, 204
pesticides 93, 216–17
Petasites spp. 161
peyote cactus 24, 65, 86, 87, 105
Phaseolus spp.:
  P. coccineus 102;
  P. lunatus 147; *P. vulgaris* 102, 146, 147, 186
pheasant's eye 58, 124
Philodendron spp. 112
Phoradendron spp. 12
phorbol esters 114, 116–19
Physostigma spp. 197;
  P. venenosum 197, 206, 207
physostigmine 197, 206–7, 213
phytoalexins 19
phytohaemagglutinin 146, 147
Picrodendron baccatum 71
picrotoxin/picrotoxinin 70, 71, 73
Pieris spp. 78
pilewort 127
pilocarpine 207
Pilocarpus spp. 197, 207;
  P. pennatifolius 207
Pimelea spp. 119;
  P. haemostachya 119
plantain family (Plantaginaceae) 54, 58, 198
Poaceae *see* grass family
podophyllotoxin 197, 209, 210–11
Podophyllum spp.:
  P. hexandrum 197, 210–11; *P. peltatum* 197, 210–11
poinsettia 21, 117
poison bulb 65, 84–5
poison hemlock 100
poison ivy/oak 109, 130–31
poison leaf 180
poison nut 66
poison parsley 100
poison pea 103, 181
polyacetylenic alcohols *see* polyynes
Polygala spp.: *P. senega* 27; *P. tenuifolia* 27
Polygonaceae *see* dock family
polyketides 25, 26, 103

polyynes (polyacetylenic acid) 64, 74, 76
pong-pong 53
poppy: Mexican prickly 173, 174–5; opium 18, 22, 27, 87, 172, 173, 202–3; pimelea 119
poppy family (Papaveraceae) 125, 157, 172–3, 174, 200
potato 42, 135, 138, 140–41
potato family (Solanaceae) 15, 27, 65, 80–82, 98–9, 122, 135, 138–9, 140, 163, 201, 217
prickly poison 178
Primula obconica 109
proscillaridin 60
protoanemonin 109, 124, 125, 126–7
Prunus spp.: *P. amygdalus* 183; *P. armeniaca* 182, 183; *P. avium* 183; *P. domestica* 183; *P. persica* 23, 183
pseudoephedrine 214, 215
Psoralea corylifolia 129
psoralens 75, 109, 128–9
Psychotria spp.; *P. ipecacuanha* 136; *P. viridis* 87
ptaquiloside 192–3
Pteridium aquilinum 179, 192–3
Pulsatilla spp. 126
purging croton 118
purging nut 114, 118
putranjiva family (Putranjivaceae) 120
pyrethrins 217
pyrethrum daisies 161, 217

Quercus petraea 33
quinine/quinidine 202, 204–5

Radix Polygalae Tenuifoliae 27
Rafflesia arnoldii 12
ragworts 161, 164, 165, 166–7, 190
Raleigh, Sir Walter 94
ramping-fumitory 172
ramsons 152
Ranunculaceae *see* buttercup family
ranunculin 124, 125, 126–7
Ranunculus spp. 124; *R. falcatus* 127; *R. ficaria* 125; *R. repens* 124; *R. testiculatus* 127
ratsbane 180, 181
rattlepod 166
Rauvolfia serpentina 69
rayless goldenrod 191
redoul 72
reserpine 69
resin spurge 116, 123
resiniferatoxin 116, 123
Retama spp. 104
Rhamnaceae *see* buckthorn family
Rhamnus spp. 144; *R. frangula* 135
Rheum spp. 144, 145; *R. hybridum* 145; *R. officinale* 144; *R. palmatum* 144, 145; *R. rhabarbarum* 145
Rhododendron spp. 21, 79; *R. ponticum* 78–9
rhubarb 144–5
Rhus spp. 130; *R. radicans* 130; *R. toxicodendron* 130; *R. typhina* 130
ricin 114, 135, 146, 147, 148–9
Ricinus communis 114, 146, 148–9
risus sardonicus 67, 77
rivastigmine 207, 213
Rivea corymbosa 89
rosary pea 146
rose family (Rosaceae) 49, 183, 202
rotenone 217
rubber 27, 52, 115
Rubiaceae *see* coffee family
rue: common 109, 129; Syrian 86
Ruta graveolens 109, 129
Rutaceae *see* citrus family

Salicaceae *see* willow family
salicin 202–3
salicylic acid 202–3
Salix spp. 202; *S. alba* 202, 203
Sanguinaria canadensis 172, 173
sanguinarine 157, 173, 174–5
Sapindaceae *see* soapberry family
saponins 27
Scadoxus spp. 142
Schinus terebinthifolius 131
Schisandraceae 70
Scilla spp. 61: *S. bifolia* 61; *S. maritima* 60, 61
scilliroside 60
scopolamine (hyoscine) 43, 65, 80, 138–9; hydrobromide 139
Scoville, Wilbur 123
Scoville heat units 116, 123
Secale cereale 89
Senecio spp. 161; *S. burchelli* 166; *S. ilicifolius* 166; *S. inaequidens* 167; *S. vulgaris* 167
senegin 27
Senna spp. 102, 144; *S. alexandrina* 102
serotonin 38, 39, 65, 110
shikimi 70, 71
Sida spp.: *S. carpinifolia* 191; *S. rhombifolia* 191
Silybum marianum 161
simplexin 119
Sinapis alba 121
Sinopodophyllum spp. 210; *S. hexandrum* 210
snake gourd 150, 151
snakeroot: Chinese 168; Indian 69; Seneca 27; white 161, 191
snowdrops 142, 212, 213
snowflakes 212
soapberry family (Sapindaceae) 158–9
Socrates 75
Solanaceae *see* potato family
solanine 140
Solanum spp. 135, 138, 139, 140; *S. dulcamara* 139; *S. elaeagnifolium* 135; *S. lycopersicum* 138; *S. melongena* 138; *S. nigrum* 138, 139; *S. scabrum* 139; *S. tuberosum* 138, 140–41
Sophora spp. 104; *S. secundiflora* 104
soporific sponge 82, 201
Sorghum spp.: *S. bicolor* 183; *S. vulgare* 183
soyabean 42
Spartium spp. 104
Spathiphyllum spp. 112
spider lilies 142, 143, 213
spider tresses 57
Spiraea spp.: *S. japonica* 49; *S. ulmaria* 202
spurge family (Euphorbiaceae) 109, 111, 114–15, 116–18, 119, 146, 148, 184
spurges 18, 114, 116–17, 123
squills 60, 61
St Ignatius bean 66
St John's wort 129
stag's horn sumac 130
star anise 64, 70, 71
Stephania tetrandra 169
steroids 25, 27, 46, 47, 52, 54–5, 96, 151, 199
stinging hairs 17, 108, 110–11, 114
Stone, Reverend Edward 202
Striga spp. 12
strophanthins 55, 56–7
Strophanthus spp. 55, 56–7; *S. amboensis* 52; *S. gratus* 57; *S. hispidus* 57; *S. kombe* 56, 57; *S. sarmentosus* 57
strychnine 25, 64, 66–7, 71, 76, 77, 96, 98, 100
Strychnos spp. 66, 67, 94, 96; *S. icaja* 67; *S. ignatii* 66; *S. nux-vomica* 64, 66, 96; *S. spinosa* 67; *S. toxifera* 96
Sturt's desert pea 179
suicide tree 53, 58
sumacs: poison 130; stag's horn 130
sunflower family *see* daisy family
Swainsona spp. 190; *S. formosa* 179
swainsonine 179, 190–91
sweet grass 171
sweet pea 187
sweet woodruff 171
sweet wormwood 205
symphytine 165
Symphytum spp. 165, 190; *S. asperum* 165; *S. officinale* 157, 164, 165; *S. x uplandicum* 165

Tabernanthe iboga 65, 68
Tanacetum spp.: *T. cinerariifolium* 161, 217; *T. coccineum* 161, 217
tanner's bush/brush family (Coriariaceae) 72
tannins 19, 23, 26, 27, 33
taro 113
taxines 46, 50, 51
Taxus spp. 46, 50, 214; *T. baccata* 46, 50–51; *T. brevifolia* 43, 46, 209; *T. jurassica* 50
teniposide 211
terpenes 25, 27, 55, 70, 126, 161, 162, 192, 193; diterpenes: esters 109, 116–19; glycosides 161; kaurene glycosides 139, 162–3; sesquiterpenes 64, 72, 73: lactones 64, 70–73; picrotoxane–type 72–3; triterpenes 47, 54, 55, 151, 216: limonoids 216–17; saponins 27
Thevetia peruviana 58
thevetins 58
thiaminase 192, 193
thistle: birdlime 157, 161, 162; bull 161; glue 162; yellow 174
thornapple 83, 138
Thymelaeaceae *see* mezereum family
tobaccos 43, 92, 93, 98–9, 139, 217
tonka beans 170
Toxicodendron spp. 109, 130; *T. pubescens* 130; *T. radicans* 130, 131
toxiferine 96
Tragia spp. 111, 114; *T. ramosa* 111
traveller's joy 125
tread carefully/softly 111

tree of the settlers 111
tremetone/tremetol 161, 179, 191
Trichosanthes cucumerina 150, 151
trillium family (Melanthiaceae) 193
tropine 80
tryptamines 38, 65, 66, 86, 87
tubocurarine 92, 95, 96
tung oil/tree 115
Turbina spp.: *T. cordata* 191; *T. corymbosa* 89
turmus 105
Tussilago farfara 161
tutin 71, 72–3
tutus 64, 72–3

Ulex spp. 104
Ungernia victoris 213
upas tree 59
Urtica spp. 110, 111; *U. dioica* 110; *U. gigas* 111; *U. gracilis* 110; '*U. urentissima* of Blume' 111; *U. spathulata* sic 111
Urticaceae *see* nettle family
urushiol 109, 130–31

van Gogh, Vincent 55
Veratrum spp.: *V. album* 43; *V. californicum* 193
Vernicia fordii 115
vilca tree 38
vincristine 208, 209
viper's bugloss 165
Viscum spp. 12

Wallace, Alfred Russel 16
warfarin 171
water dropworts 74, 76–7
water hemlocks 64, 76–7
Wedelia glauca 163
willow 202, 203
willow family (Salicaceae) 202
wintergreen 203
witchweeds 12
Withering, William 198
wolfsbane 14, 48

Xanthium spp. 139; *X. strumarium* 156, 162, 163
Xanthosoma sagittifolium 113
xi shu 43

yew 46, 50–51, 214; Pacific 43, 46, 209, 211
yohimbe 69
yopo 65, 86, 102

Zamia furfuracea 189
Zamiaceae 188, 189
zephyr lilies 142
Zephyranthes spp. 142
Zubrówka 171

Acknowledgements

The authors would like to thank the fantastic editorial team, Jacqui, David and Susi, and designer Lindsey for their support throughout the process of writing and producing this book.

Liz is grateful to Dr Christine Leon MBE and Professor Virginia Murray who introduced her to poisonous plants when they worked for Guy's Poisons Unit, and to her many colleagues both at Guy's & St Thomas' Hospital, London, and the Royal Botanic Gardens, Kew, who share her love of plants and interest in both their useful and darker sides; thanks go to Dr David Goyder for identification of Apocynaceae images. Lastly, it would not have been possible to write this book without the support and understanding of her family.

Sonny acknowledges that this book would not have been written without the constant understanding and encouragement from Liz! He is also grateful to past and present colleagues at the botanical gardens of Kew, Copenhagen and Uppsala, as well as at the Swedish Poison Information Centre.

Picture credits

Illustrations on pages 17, 19, 20, 25, 30 (left & right), 31, 33 (top left & right) by Robert Brandt.

All images copyright the following (T = top, M = middle, B = bottom, L = left, R = right):

Alamy Stock Photo: 5 Wildscotphotos; 9R Homer W Sykes; 21 Lucy Turnbull; 23R blickwinkel; 42T age fotostock; 43T imageBROKER; 48 Garden World Images Ltd; 51 Homer W Sykes; 56 AfriPics.com; 58L Bramwell Flora; 61 Universal Images Group North America LLC/DeAgostini; 64B Manfred Ruckszio; 69TL Universal Images Group North America LLC; 69B mauritius images GmbH; 70L John Richmond; 71 blickwinkel; 76 Anne Gilbert; 77 blickwinkel; 80 blickwinkel; 85 Armands Pharyos; 86 Brian Van Tighem; 87 Pawel Bienkowski; 105T Yon Marsh Natural History; 110 Suzanne Long; 115B Rob Matthews; 116 age fotostock; 118 Stefano Paterna; 119B Stephanie Jackson – Aust wildflower collection; 129B Medicshots; 135 blickwinkel; 136 WILDLIFE GmbH; 139 WILDLIFE GmbH; 146R Design Pics Inc; 150 Tim Gainey; 156 imageBROKER; 157T Frank Hecker; 161B WILDLIFE GmbH; 162 imageBROKER; 163B Florapix; 166 Nigel Cattlin; 169 WILDLIFE GmbH; 173B Carol Dembinsky/Dembinsky Photo Associates; 178L Garden World Images Ltd; 185B Hilke Maunder; 187T Rick & Nora Bowers; 189T Florapix; 193B Richard Griffin; 200 Chrispo; 203B blickwinkel; 205T robertharding; 209 Minden Pictures; 211 Organica; 213T Graham Prentice; 214 WILDLIFE GmbH; 217T QpicImages.
Liz Dauncey: 74B.
John Grimshaw: 16R.
Nature Photographers Ltd: 9L Laurie Campbell; 14R Laurie Campbell; 50 Laurie Campbell; 100 Paul Sterry; 101T Paul Sterry; 128 E.A. Janes; 170 Paul Sterry; 172BR Paul Sterry; 192 Paul Sterry.
MedicalArtist.com: 46L.
Damir Repič: 108B.
Science Museum, London, Wellcome Images: 196R, 205B.
Science Photo Library: 25T Biology Pics; 41M & B Trevor Clifford Photography; 108T Antonio Romero; 109T Sheila Terry; 197T Jerry Mason.
Shutterstock: 1, 2, 10, 28, 44, 62, 132 Geraria; 8 martaguerriero; 12 Alexander Mazurkevich; 13 AustralianCamera; 15 Martin Fowler; 16L Andrew Koturanov; 17T Ken Wolter; 18 BMJ; 19T D. Kucharski K. Kucharska; 20 plamice; 22T vseb; 22B Dr Morley Read; 23L Only Fabrizio; 24 vovan; 27 marilyn barbone; 33 Manfred Ruckszio; 34 Andrea Danti; 35BL Potapov Alexander; 35BR Oliver S; 36 Alila Medical Media; 37T siriboon; 37B Vector FX; 38 Somrerk Witthayanant; 41T Ariene Studio; 26 tristan tan; 42B Kelly Marken; 43B Catalin Petolea; 46R Manfred Ruckszio; 47T Volkova Irina; 47B Katsiuba Volha; 49 Alex Polo; 53TL kongsky; 53BL Porawas Tha; 53BR Cathy Keifer; 54 Morten Normann Almeland; 57T Panya7; 58R Boonchuay1970; 59B Alexlky; 64T Morphart Creation; 65T Shulevskyy Volodymyr; 65B Ammit Jack; 66 wasanajai; 67 Manfred Ruckszio; 70R Dolores Giraldez Alonso; 72 Wiert nieuman; 78 vagabond54; 79 Dennis van de Water; 81 A_lya; 82B KBel; 83B LFRabanedo; 88 Maljalen; 89 Carmen Rieb; 90 Morphart Creation; 92B Anne Kitzman; 93T Robert Biedermann; 93B Michael Avory; 95 Ammit Jack; 97T Catchlight Lens; 97B Bill Perry; 98 Dariusz Leszczynski; 102B topimages; 104 MaryAnne Campbell; 105B Olga Popova; 109M wjarek; 109B ncristian; 112 Niwat Sripoomsawatt; 113T the808; 113B SOMMAI; 114B joloei; 115B Kidsada Manchinda; 117 SAPhotog; 119T Popova Valeriya; 120 Bildagentur Zoonar GmbH; 121T Patana; 121B Madlen; 122 photo one; 123T mangbiz; 123B stolekg; 124BL Mariola Anna S; 124BR SRichard Griffin; 126R APugach; 127 Ann Louise Hagevi; 129T annalisa e marina durante; 130 Flashon Studio; 131T Tom Grundy; 131B Aggie 11; 140 Irina Borsuchenko; 141 Pixeljoy; 142L Annaev; 142R EM Arts; 143T FlorinRO; 143B Yayuyu210615; 144 Lotus Images; 145T photoiconix; 145B Manfred Ruckszio; 146L Caner Cakir; 147BL jeehyun; 147BR Lotus Images; 148 Xico Putini; 149L Dimijana; 149R Ruttawee Jai; 151 Thongseedary; 152 Wolfgang Simlinger; 153T nofilm2011; 157B SeDmi; 158 twiggyjamaica; 159 apiguide; 160B oksana2010; 161T Nancy Bauer; 164 Luka Hercigonja; 165 BergeImLicht; 167 Ruud Morijn Photographer; 168 Martin Fowler; 171L Nattika; 171R fotomarekka; 172BL LesPalenik; 174 Ovchinnikova Irina; 175 StripedNadin; 176 Libellule; 179B Ashley Whitworth; 181B Ricardo de Paula Ferreira; 182 petrovichlili; 183L JIANG HONGYAN; 184 Yatra; 185T Fecundap stock; 186 ARCANGELO; 188 JT888; 189B Antonio Gravante; 190 cjchiker; 191T ALong; 191B Wiert nieuman; 193T Steve Cukrov; 194 Morphart Creation; 196L Morphart Creation; 197B Birute Vijeikiene; 199 Alexander Varbenov; 201T Andrew Koturanov; 202 Kalcutta; 208 chanwangrong; 212 Elena Koromyslova; 215B marilyn barbone; 216 Antony R.
Wellcome Library, London: 17M, 94, 198L, 215T.
Wikimedia Commons: 52B Ikiwaner; 57B Maša Sinreih in Valentina Vivod; 59T Wibowo Djatmiko; 68 Marco Schmidt; 73 MurielBendel; 83T Didier Descouens; 99 Dcrjsr; 101B the Providence Lithograph Company; 111T George Yatskievych; 111B Stan Shebs; 126L H. Zell; 134 Stan Shebs; 137B Maša Sinreih in Valentina Vivod; 179T John Hill; 180 JMK; 181T JMK; 183R INRA DIST.

All other images in this book are in the public domain.

Every effort has been made to credit the copyright holders of the images used in this book. We apologize for any unintentional omissions or errors, and will insert the appropriate acknowledgement to any companies or individuals in subsequent editions of the work.